Design of New Weave Patterns

Design of New Weave Patterns

Radostina A. Angelova

CRC Press
Taylor & Francis Group
Boca Raton London New York

CRC Press is an imprint of the
Taylor & Francis Group, an **informa** business

CRC Press
Taylor & Francis Group
6000 Broken Sound Parkway NW, Suite 300
Boca Raton, FL 33487-2742

Printed on acid-free paper

International Standard Book Number-13: 978-0-367-34785-7 (Hardback)

Library of Congress Cataloging-in-Publication Data

Names: Angelova, Radostina A., author.
Title: Design of new weave patterns / Radostina A. Angelova.
Description: First edition. | New York, N.Y. : CRC Press/Taylor & Francis Group, 2019. | Includes bibliographical references and index.
Identifiers: LCCN 2019032834 (print) | LCCN 2019032835 (ebook) | ISBN 9780367347857 (hardback ; acid-free paper) | ISBN 9780429327971 (ebook)
Subjects: LCSH: Weaving--Patterns.
Classification: LCC TS1490 .A536 2019 (print) | LCC TS1490 (ebook) | DDC 677/.028242--dc23
LC record available at https://lccn.loc.gov/2019032834
LC ebook record available at https://lccn.loc.gov/2019032835

Visit the Taylor & Francis Web site at
http://www.taylorandfrancis.com

and the CRC Press Web site at
http://www.crcpress.com

Contents

SECTION II New Methods for Design of New Weave Patterns

Foreword

The structure of the woven textiles is one of the essential features that make them different from other textiles and materials. The visual appearance, handle, texture, and design possibilities are extremely preconditioned by the weave pattern.

The creation of a new weave pattern is a challenging task. Though the number of patterns for textiles seems to be illimitable, only three basic patterns stay on the bottom of the whole diversity of woven fabrics with different applications: for clothing, interior textiles, textiles in means of transport, textiles for protective clothing, or technical textiles. These are the elementary weaves: the plain weave, twill, and satin/sateen. All other patterns are based on the elementary weaves, and it is often difficult for the designers to create a new weave, which is adequately related to the application, can meet the requirements of the customer, and is free of a pattern patent (Grosicki, 2004, Angelova, 2005).

Despite the hundreds and thousands of weave patterns, developed within the ages, the number of methods for their creations is limited (Milasius & Reklaitis, 1988). These methods involve (Angelova, 2017):

- *Mathematical method.* Perhaps this is the most frequently used method for the design of a new weave pattern. The basic approaches involve operations with images: rotation, inversion, and mirror images. Mathematical expressions, based on the dependences between warp and weft interlacing, have also been applied: Shinohara et al. (2008) performed automatic construction of weave patterns using the positioning of the yarn in the fabric. Basics from crystallography were also applied for the development of new weave patterns (Washburn, 1988). Xiao et al. (2015) presented a study on geometrical modeling of honeycomb woven fabrics. New weave patterns of composite structures are also frequently researched (Adumitroaie & Barbero, 2011, Dash & Behera, 2018).
- *Pattern method.* It is based on the creation of a motif, which is filled up with already known patterns. Combination of motifs can also be arranged to make the new weave pattern complex or unique (Washburn, 1988, Grosicki, 2004).
- *Historical method.* Cultural traditions in the weaving of specific textile items, costumes, or interior textiles are also a source of inspiration for new patterns development. The method could also involve imitations of patterns from other handicrafts, i.e., embroidery or knitting (Huang & Zhou, 2018). Many researchers have found inspiration in the design of garments, typical for different regions and cultures. Watt (1998) has explored the patterns in the design of woven textiles in Italy from the seventeenth century. Arthur (2008, 2011) dedicated her efforts to the design of Hawaiian quilting, while Rikert et al. (1999) investigated Swedish patchwork quilts.

- *Random method.* It could be widely used, especially in stages of "deep inspiration" of the creator. This specific technique involves adding or removing of warp and weft overlaps within a limited pattern, which dimensions (warp and weft repeat) are already set. In this method, automatic generators of binary codes or systems working with matrices can replace the human activity (Chen et al. 1996). The disturbance of warp and weft overlapping in already known patterns is also a way to create a new pattern from an existing one.

REFERENCES

Adumitroaie, A., & Barbero, E. J. (2011). Beyond plain weave fabrics – I. Geometrical model. *Composite Structures, 93*, 1424–1432.

Angelova, R. A. (2005). Designing new weaving patterns using Boolean operations. *Proceedings from X International Conference of the Faculty of Power Engineering, EMF'2005* (Vol. 2, pp. 137–142), Sozopol, Bulgaria.

Angelova, R. A. (2017). Design of weave patterns: When engineering textiles meets music. *The Journal of the Textile Institute, 108*(5), 870–876.

Arthur, L. (2008). The influence of Hawaii's tropical environment on Hawaiian quilting design. *Journal of Interdisciplinary and Cross Cultural Studies, 4*(1), 1–21.

Chen, X., Knox, R. T., McKenna, D. F., & Mather, R. R. (1996). Automatic generation of weaves for the CAM of 2D and 3D woven textile structures. *Journal of the Textile Institute, 87*(2), 356–370.

Dash, A. K., & Behera, B. K. (2018). Role of weave design on the mechanical properties of 3D woven fabrics as reinforcements for structural composites. *The Journal of the Textile Institute, 109*(7), 952–960.

Grosicki, Z. J. (2004). *Watson's Advanced Textile Design, Compound Woven Structures*, 4th Ed. Reprinted by Woodhead Publishing Ltd. Cambridge, UK.

Huang, J., & Zhou, X. (2018, June). Oroqen decorative pattern design and research on the application of creation thought. In *IOP Conference Series: Materials Science and Engineering* (Vol. 371, No. 1, p. 012062). IOP Publishing.

Milasius, V., & Reklaitis, V. (1988). The principles of weave-coding. *Journal of the Textile Institute, 79*(4), 598–605.

Rikert, S. L., Harp, S. S., Horridge, P. E., & Shroyer, J. L. (1999). Documentation of Swedish patchwork quilts: 1830 to 1929. *Clothing and Textiles Research Journal, 17*(3), 134–143.

Shinohara, T., Takayama, J.-Y., Ohyama, S., & Kobayashi, A. (2008). Automatic weave diagram construction from yarn positional data of woven fabric. *Textile Research Journal, 78*, 745–751.

Washburn, D. K. (1988). Analysis of pattern structure by geometric symmetries. *Textiles as Primary Sources: Proceedings of the First Symposium of the Textile Society of America*, Minneapolis Institute of Art, September 16–18.

Watt, M. (1998). Exploring pattern in woven design: A comparison of two seventeenth century Italian textiles. *Proceedings from Textile Society of America Symposium*, 446–455, New York.

Xiao, X., Hua, T., Li, L., & Wang, J. (2015). Geometrical modeling of honeycomb woven fabric architecture. *Textile Research Journal, 85*, 1651–1665.

Preface

The design process in engineering requires modification, so as engineers and designers to efficiently tackle the demands of the markets and society. Clothing and textiles industries are constantly pressured by the market and consumers to offer more diverse products. The market pressure affects designers of fabrics particularly, as they have to meet the requirements of both haute couture and average consumers, who tend to spend less money for clothing and textiles but require higher quality and originality in terms of fashion trends.

The monograph *Design of New Weave Patterns* presents a systematic study on methods used for the creation of weave patterns for simple structures. The design of new weaves is spread over fabrics, involving one system of warp threads and one system of weft threads that intersect one another at right angles.

The first section of the monograph is dedicated to the classical methods for the design of new weave patterns. A brief history of the weave patterns is presented in Chapter 1. The construction of the woven fabrics and the specifics of the weave patterns, as well as their characteristics, are presented in Chapter 2. The elementary weaves (the plain weave, twill, and satin/sateen) and their derivatives are described in Chapters 2 and 3.

A merit of the first section of the monograph is the presentation of a new classification of the techniques used at present for designing of new weave patterns. Four different methods are summarized and discussed in detail, being largely supported by figures of patterns, color view, and simulations of the fabric's appearance:

- Design of new weave patterns based on patterns merge (Chapter 4). Weave patterns for vertical and horizontal stripe effects and cell-like effects are presented.
- Design of new weave patterns based on motifs (Chapter 5). The method allows designing of cloths with cell-like effects, diagonal effects, and zig-zag effects, as well as crepe fabrics.
- Design of new weave patterns based on patterns insertion (Chapter 6). The discussed method is applied for the design of rib-like effects, diagonal effects, and crepe weaves.
- Design of new weave patterns based on a change of the displacement number (Chapter 7). The change of the displacement number allows the design of new weave patterns for diagonal effect, zig-zag effects, and curved effects.

The second section of the monograph proposes new, original methods for the design of new weave patterns:

- New weave patterns based on Boolean operations (Chapter 8). The method has been proposed in the work of Griswold (2002). It is based on mathematical operations with matrices, and new aspects of the method are discussed.

- New weave patterns based on musical scores (Chapter 9). The method is based on the visual analogy between the musical scores and the weave diagrams. It has already been presented to the research community with plenty of color design solutions (Angelova, 2017).
- New weave patterns based on text interpretations (Chapter 10). The method is based on the transition from letters to numbers using different transformation matrices. It is presented for the first time in the book.
- New weave patterns based on the Braille alphabet (Chapter 11). The method uses both the image of the Braille cells and the dots in each cell. It is also presented in the monograph for the first time.

All design methods in Section II are supported by an explanation of the performed transformations. The advantages and disadvantages of the methods are also discussed, together with the assessment of the new weave patterns. The possible applications are commented, and several woven fabric designs are used for visualization.

Acknowledgments

I dedicate this monograph to my parents and brother. My mother Vanya has always been my best guide in the world of beauty and excellent style. My father Angel taught me to engage people with myself only when I really have something to tell them. Also, my brother Iliyan is among the best engineers I know and the best brother I could beg for.

The weave patterns are one of my passions. The creation of a new weave pattern and woven textiles is one of the best and enjoyable cross sections between art and engineering science.

I am thankful to *The Journal of the Textile Institute* (https://www.tandfonline.com/) for their permission to use parts of my article (Angelova, RA, 2017, *The Journal of the Textile Institute*, 108(5), 870–876) in this monograph.

I am grateful to my students, who are an incentive for me to find answers.

Last but not least, I thank my husband Prof. Peter Stankov and our son Alexander for their love and patience.

As always, I am grateful to God for the inspiration and everything that has led to the creation of this book.

Radostina A. Angelova
Sofia, 2019

Author

Radostina A. Angelova is an associated professor at the Department of Textiles, Technical University of Sofia (TU-Sofia). She is also an associated researcher at the Centre for Research and Design in Human Comfort, Energy and Environment at TU-Sofia. She is a lecturer in spinning technologies and machines, weaving technologies and machines, and design of yarns and fabrics. Angelova holds an MSc degree in textiles and clothing (1994), a PhD in technology of textile materials (2001), and a DSc in technology of textile materials (2015). She authored the book *Textiles and Human Thermophysiological Comfort in the Indoor Environment* (CRC Press, 2015), more than 150 papers, 4 textbooks, and 8 chapters in edited books.

Section I

Classical Methods for Design of New Weave Patterns

1 Brief History of Weave Patterns

The woven fabrics are 3D macrostructures produced due to the interlacing of at least two sets of threads: vertical (warps) and horizontal (wefts). Usually, the woven fabric is produced on a weaving machine (loom). Figure 1.1 shows a picture of a handloom, and Figure 1.2 shows a modern air-jet jacquard weaving machine.

The macrostructure of the woven fabric consists of two levels (Figure 1.3):

- Mesostructure, formed by the threads: yarns of filaments;
- Microstructure, consisting of fibers that are spun into yarns or monofilaments that constitute the filaments (polyfilaments).

The Latin word *textilis* means *woven*. The modern term *textiles* is frequently associated with woven fabrics, but nowadays the textiles can have three different types of a macrostructure: woven, knitted, and nonwoven.

The most accessible and probably first used by the humans fibers for textiles were extracted from lime tree, oak, willow, or elm. Among them non-fibrous raw materials were also used, applying the same principles of interlacing flexible branches or reed. Flax is considered to be the earliest domesticated plant, followed by hemp: both of them associated with Asia as a homeland. Kvavadze et al. (2009) reported the discovery of wild flax dyed fibers, used in a fabric, in the Dzudzuana Cave in Georgia, dated from 32,000 to 26,000 years BP. Though the appearance of some disagreements (Bergfjord et al., 2010), these flax fibers are thought to be the oldest evidence of textiles used by humans.

In Europe, the oldest artifact for textiles/basketry use dates back 25,000–27,000 BP, discovered in Dolní Vestonice I, Czech Republic (Soffer et al., 2001). The next oldest proof of fiber-based technology developed by humans came from the Sea of Galilee in Israel, Ohalo II: Z-twist cordage from an unknown plant species was dated to be produced 19,300 years BP (Nadel et al., 1994).

The oldest evidence for textiles production in Asia comes from the Far East (Derevyanko & Medvedev, 1995) and dates to 13,500 BP. In North America, the oldest textiles artifacts are dated to be produced around 11,000 years BP (Andrews & Adovasio, 1996) and in South America, in Monte Verde, Chile, dated around 13,000 BP (Adovasio, 1997).

The oldest weave pattern, used by humans, is the plain weave. Long before the manufacturing of the first textile, it was applied for basketry production. In the plain weave, the first weft thread goes alternately under the first warp thread, then over the second warp thread, etc. The second weft thread goes first over the first warp thread, then under the second warp thread, etc. Thus, a stable macrostructure is formed.

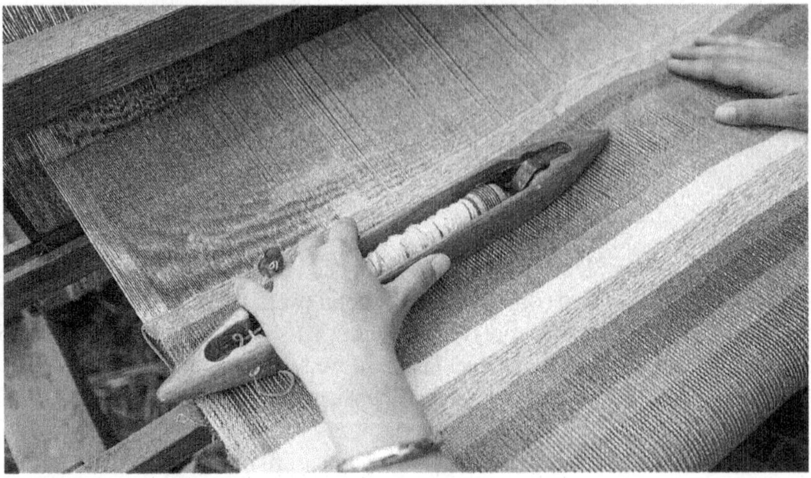

FIGURE 1.1 Handloom.

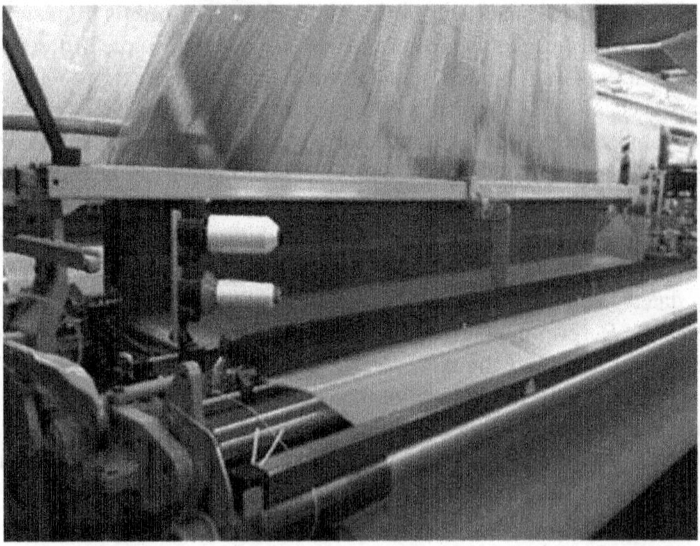

FIGURE 1.2 Air-jet jacquard weaving machine.

Figure 1.4 shows a microscopic picture of cotton fabric in a plain weave. Popular fabrics in this weave are percale, canvas (Figure 1.5), organza, chiffon, and taffeta.

The twill weave breaks the alternation of the plain weave, as not every third weft thread repeats the first (and every fourth weft thread repeats the second), but at least every fourth. The threads interlacing is more complicated and leads to the formation of diagonal ribs. The two sides of the woven fabric look differently; one of them is considered to be the *face* of the fabric (usually that on the upper side of the weaving loom) and the other the *back* of the fabric.

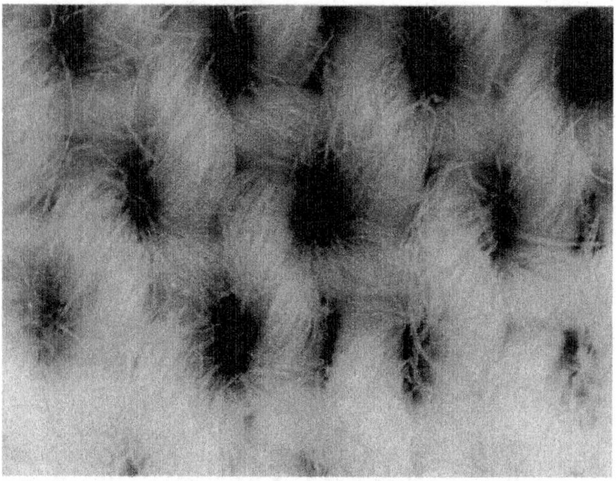

FIGURE 1.3 Microscopic view of a woven textile macrostructure with visible fibers (microstructure) spun into yarns (mesostructure).

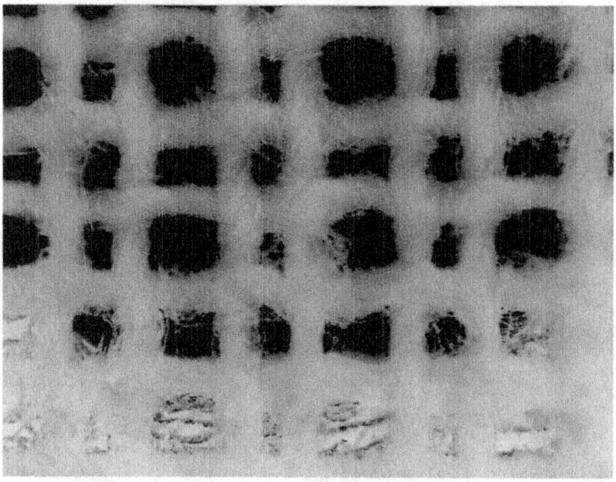

FIGURE 1.4 Microscopic view of woven textile in a plain weave.

Figure 1.6 shows a microscopic view of cotton fabric in a twill weave, with visible left-hand diagonals. Denim is the most popular modern fabric in a twill weave (Figure 1.7). Other examples for textiles in twill weaves are gabardine, chino, tweed, and serge.

Satin or sateen are weaves that lead to the production of fabrics with a glossy surface that reflects perfectly the light. The difference between satin weave and sateen weave is discussed in Chapter 2. The application of filaments instead of spun yarns increases the glossy effect. The name of the weave pattern comes from the port city of Quanzhou (China), pronounced in medieval Arabic as "Zayton" (or Zaytun) (So & Su, 2000).

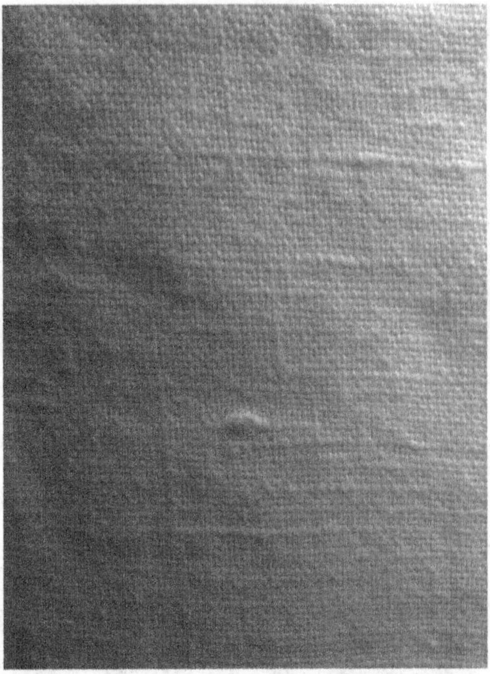

FIGURE 1.5 Canvas.

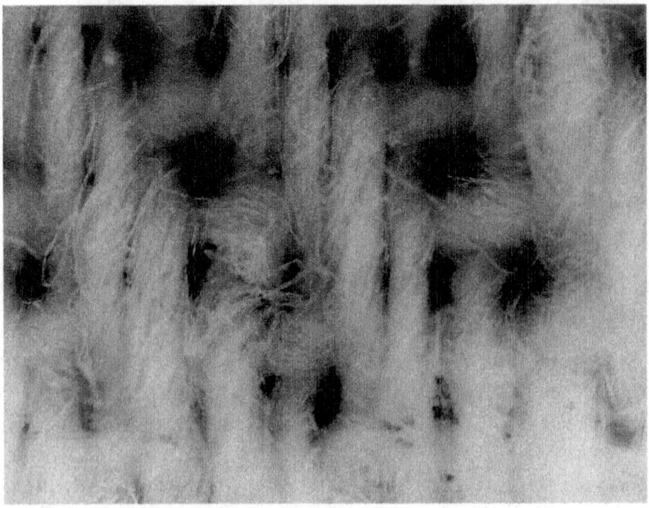

FIGURE 1.6 Microscopic view of woven textile in a twill weave.

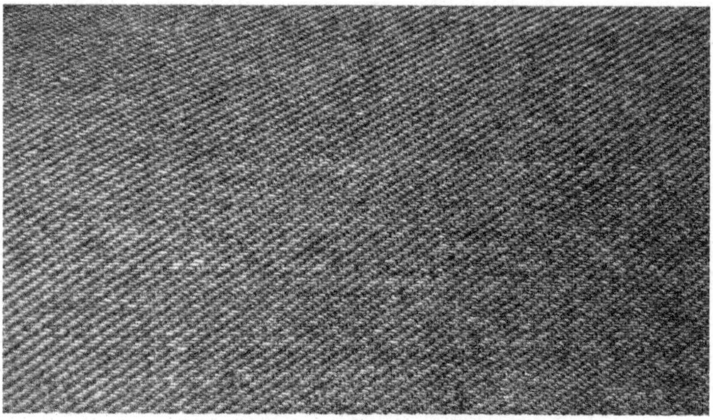

FIGURE 1.7 Denim.

Figures 1.8 and 1.9 show a microscopic picture and the view of fabric in a satin weave, respectively.

All other weave patterns appeared from the first three elementary weaves. Nowadays, the following groups of weave patterns can be distinguished:

- Elementary weaves—plain weave, twill, and satin/sateen. They appear at a time when weaving is manual, slow, and cannot weave patterns that require complex movement of the warp threads.
- Derivatives of the elementary weaves. They are, as a rule, with small repeats and can be made on hand looms or weaving machines with a small number of harnesses. Exceptions can be shadowed twill weave and shadowed sateen weave (presented in Chapter 3).
- Compound weaves that are a combination of two or more elementary weaves, their derivatives, or already known compound weaves. They appear with the development of the looms' construction and increment of the production speed, although some of them can still be made with little difficulty in handlooms or weaving machines with a small number of harnesses.
- Complex weaves, for which three or more sets of threads are needed (e.g., the production of terry fabrics requires two sets of warp threads and one set of weft threads). They appear late because they require unique mechanisms to control more than one system of warp threads or mechanisms to insert weft threads in the shed with complex formation.
- Jacquard weaves that are based on the application of jacquard weaving machines and require individual motion of warp threads. The process of the Jacquard weaves appearance was inevitably linked to the invention of the Jacquard loom in 1804, where a continuous chain of punched cards controlled the warp threads. Today, it is considered a discovery that led to the emergence of binary code and computer technology.

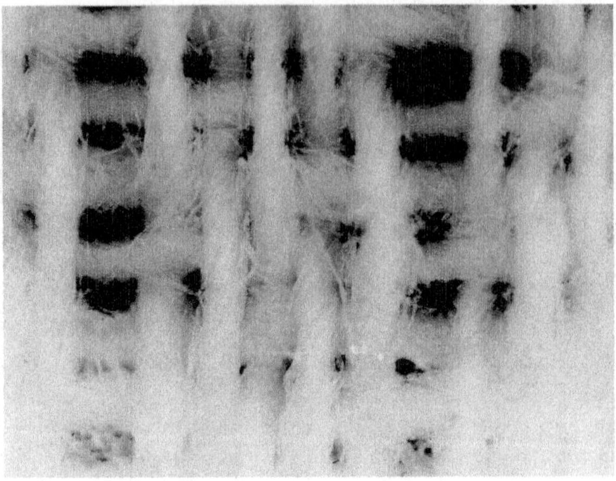

FIGURE 1.8 Microscopic view of woven textile in a satin weave.

FIGURE 1.9 Satin.

REFERENCES

Adovasio, J. M. (1997). Cordage and cordage impressions from Monte Verde. *Monte Verde. A Late Pleistocene Settlement in Chile*, 2, 221–228. Smithsonian Institution Press, Washington, DC.

Andrews, R. L., & Adovasio, J. M. (1996). The origins of fiber perishables production east of the Rockies. *A Most Indispensable Art: Native Fiber Industries from Eastern North America*, 30–49. The University of Tennessee Press, Knoxville, TN.

Bergfjord, C., Karg, S., Rast-Eicher, A., Nosch, M. L., Mannering, U., Allaby, R. G., Murphy, B. M., & Holst, B. (2010). Comment on "30,000-year-old wild flax fibers". *Science*, *328*(5986), 1634–1634.

Derevyanko, A. P., & Medvedev, V. E. (1995). The Amur River basin as one of the earliest centres of ceramics in the Far East. *The Origins of Ceramics in East Asia and the Far East.* Tohoku, 13.

Kvavadze, E., Bar-Yosef, O., Belfer-Cohen, A., Boaretto, E., Jakeli, N., Matskevich, Z., & Meshveliani, T. (2009). 30,000-year-old wild flax fibers. *Science*, *325*(5946), 1359–1359.

Nadel, D., Danin, A., Werker, E., Schick, T., Kislev, M. E., & Stewart, K. (1994). 19,000-year-old twisted fibers from Ohalo II. *Current Anthropology*, *35*(4), 451–458.

Soffer, O., Adovasio, J. M., Hyland, D. C., Illingworth, J. S., Klima, B., & Svoboda, J. (2001). Perishable industries from Dolní Vestonice I: New insights into the nature and origin of the Gravettian. *Archaeology Ethnology and Anthropology of Eurasia*, 2, 48–65.

2 Elementary Weaves

2.1 FABRIC CONSTRUCTION AND SPECIFICS OF THE WEAVE PATTERN

The construction of a fabric is determined by the mutual arrangement of the warp and weft threads, the relationship between their characteristics and number, and the interaction between them. The way of interlacing between the warp and weft threads is called a *weave*.

2.1.1 Warp and Weft Overlaps

During the formation of the shed, the warp threads can move only to two positions – to rise or to descend. In this situation, the upper warp threads are placed above the weft threads and the lower weft threads are under the weft threads. When a warp thread is at the top, a *warp overlap* is formed, and when the weft thread is at the top (the warp thread is under the weft thread) – a *weft overlap* is formed (Figure 2.1). The combination of warp and weft overlaps leads to the creation of different weaves.

For the presentation of the weave pattern on a design paper, a binary code is used: any kind of a mark in the cell, representing the intersection point between a warp and a weft thread, indicates a warp overlap. The empty cell indicates a weft overlap. Figure 2.2a shows the *2D binary code* or the *weave diagram*, used for representation of a right-handed twill 1/3: the full cell corresponds to a warp overlap (the warp yarn is above the weft yarn), and each empty cell shows a weft overlap (the weft yarn is above the warp yarn). Figure 2.2b is a pseudo-3D representation of the same weave, which gives an idea about the texture of the fabric.

2.1.2 Warp and Weft Floats

Float is the consecutive placement of a thread of one of the thread sets (warp or weft) over two or more threads of the other set (Figure 2.3). Longer the float more unstable the woven structure is, due to the lesser number of interlacing points between warp and weft threads. All weave patterns, except plain weave, involve shorter or longer float of warp and/or weft threads.

Group float is also possible, like in canvas or rib weave, which are discussed in detail in Chapter 3. Figure 2.4 illustrates the canvas weave 4/4, where four warp threads fly over four weft threads and vice versa.

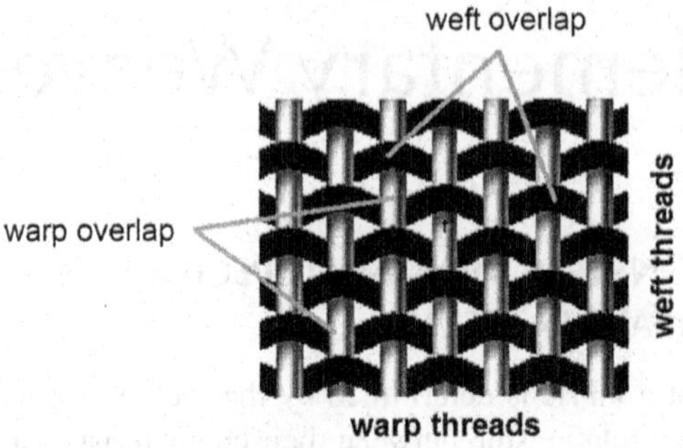

FIGURE 2.1 Structure of a fabric: sets of warp and weft threads; warp and weft overlaps.

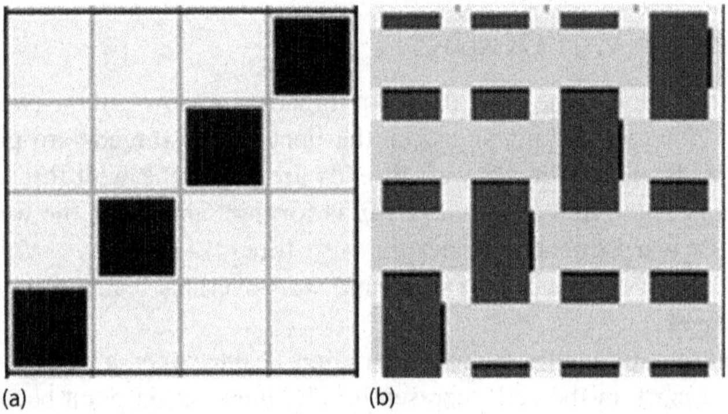

FIGURE 2.2 Weave pattern (right-handed twill 1/3): (a) weave diagram and (b) representation of the interlacing between warp and weft threads.

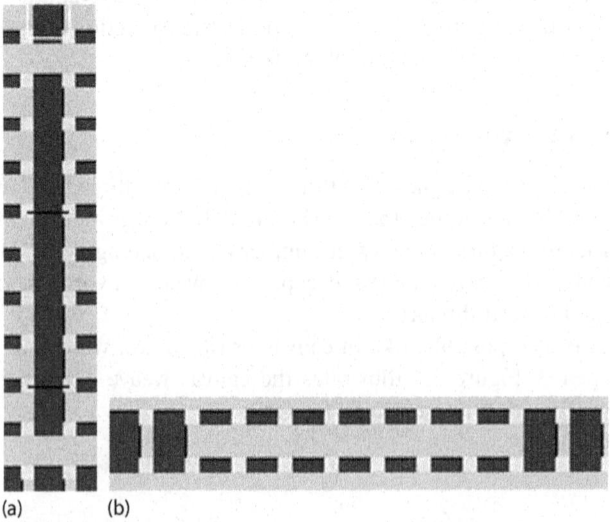

FIGURE 2.3 Thread float in the cloth structure: (a) warp float and (b) weft float.

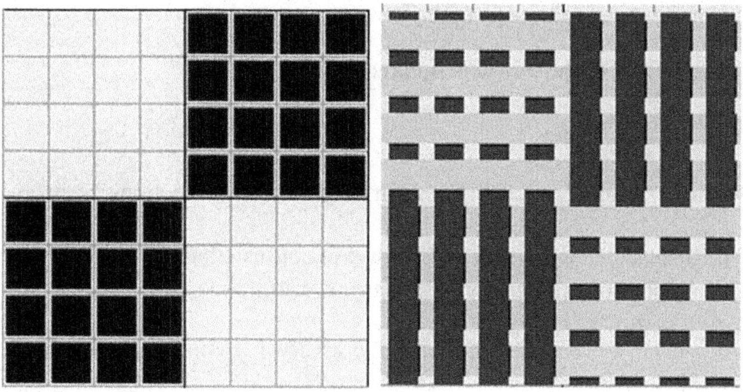

(a) (b)

FIGURE 2.4 Group floats (canvas weave 4/4): (a) weave diagram and (b) representation of the interlacing between warp and weft threads.

2.1.3 Characteristics of the Weaves

Three characteristics define each weave and can be used for their unique description. This description is especially important when two or more weave patterns from the same type are compared and assessed for advantages or disadvantages.

The weave characteristics are as follows (Damyanov & Chobanov, 1988):

- *Repeat* (*R*): the smallest number of warp and weft overlaps that define the whole cloth.
- *Number of overlaps* (*n*): the total number of warp n_{wp} and/or weft n_{wf} overlaps in the weave repeat. When $n_{wp} > n_{wf}$, the result is a *warp-faced* fabric. When $n_{wf} > n_{wp}$, the fabric is with a *weft-faced* view.
- *Displacement number* (*S*): it shows how the next overlap in the repeat is situated compared to the previous one. The displacement number is an integer with a different sign (plus or minus):
  - When a warp thread is considered, S_{wp} is positive if the overlap shifts upward and S_{wp} is negative if the overlap shifts downward from the previous overlap.
  - When a weft thread is considered, S_{wf} is positive if the overlap shifts rightward and S_{wf} is negative if the overlap shifts leftward from the previous overlap.

2.2 ELEMENTARY WEAVES

The number of weave patterns is uncountable. However, all they are based on three *elementary weaves*: plain weave, twill weave, and sateen/satin weave.

The main features of the *elementary weaves* are:

- The repeat R of the weave is square:

$$R = R_{wp} = R_{wf} \quad (2.1)$$

where R_{wp} and R_{wf} are the repeats in warp and weft directions, respectively.

- The displacement number S of the weave construction is constant.
- The number of overlaps for each warp or weft thread is:

$$n_{wp} = 1 \text{ and } n_{wf} = R - 1 \quad (2.2)$$

or

$$n_{wp} = R - 1 \text{ and } n_{wf} = 1 \quad (2.3)$$

where n_{wp} and n_{wf} are the warp and weft overlaps alongside the yarn in the repeat of the weave.

2.2.1 Plain Weave

The plain weave is the oldest weave and has the smallest repeat $R = 2$ (Figure 2.5). There are two warp overlaps, and two weft overlaps in the repeat. The resulting cloth is stronger and firmer than any other fabric.

The plain weave has two equal faces and one of them is considered as a technical face. It is probably the most used weave in the world, as it allows a great variety of fabric views and textures, depending on the threads and their material, thickness, color, and twist.

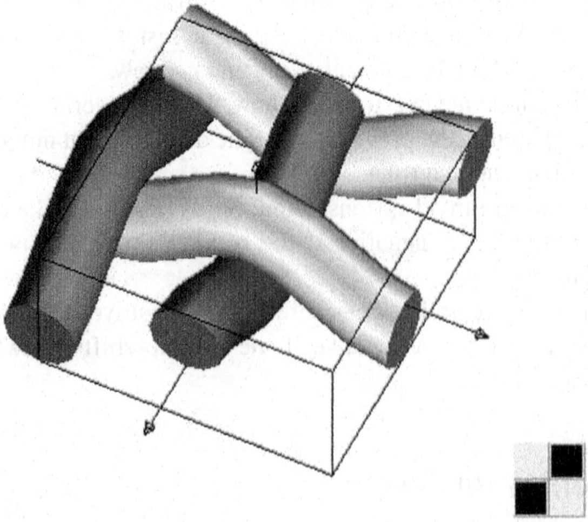

FIGURE 2.5 Plain weave.

2.2.2 Twill Weave

The weaving effect of the twill is diagonals with different width and face (warp-faced and weft-faced diagonals). The repeat of the simple twill is $R = R_{wp} = R_{wt} \geq 3$.

In the repeat, the elementary twill has one warp overlap alongside a thread, and the rest are weft overlaps or vice versa: one weft overlap is followed by warp overlaps. The following expressions are valid for each warp or weft thread:

$$n_{wp} = 1 \text{ and } n_{wf} = R - 1 \geq 2 \tag{2.4}$$

or

$$n_{wp} = R - 1 \geq 2 \text{ and } n_{wf} = 1 \tag{2.5}$$

Figure 2.6 shows right-handed twill (or Z-twill) $\frac{1}{2}Z$ with only one warp overlap alongside a single yarn; the rest are weft overlaps. The same is valid for the right-handed twill $\frac{2}{1}Z$ (Figure 2.7), but the single overlap is a weft one, and the rest are warp overlaps.

The diagonal effect comes from the displacement of the single overlap in a horizontal and vertical direction with one thread in relation to the previous single overlap. The displacement number is $S = \pm 1$, which results in a different direction of the diagonals.

Right-handed twill (Figure 2.8) is obtained when the following combination of displacement numbers is applied:

$$S_{wp} = +1 \text{ and } S_{wf} = +1 \tag{2.6}$$

or

$$S_{wp} = -1 \text{ and } S_{wf} = -1 \tag{2.7}$$

where S_{wp} is the displacement number in the direction of the warp threads and S_{wt} is the displacement number in the direction of the weft threads.

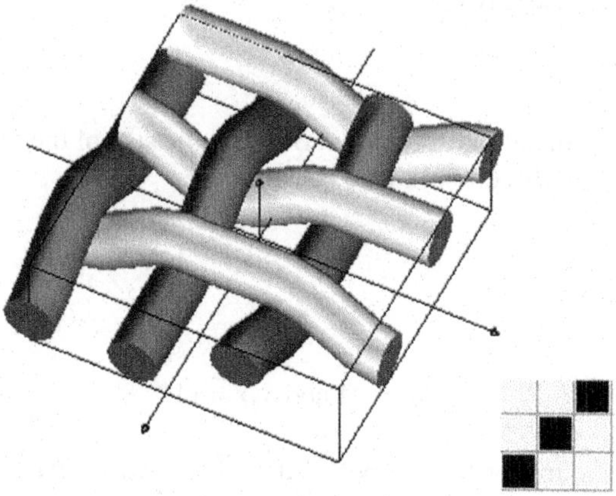

FIGURE 2.6 Twill weave $\frac{1}{2}Z$ (weft-faced pattern).

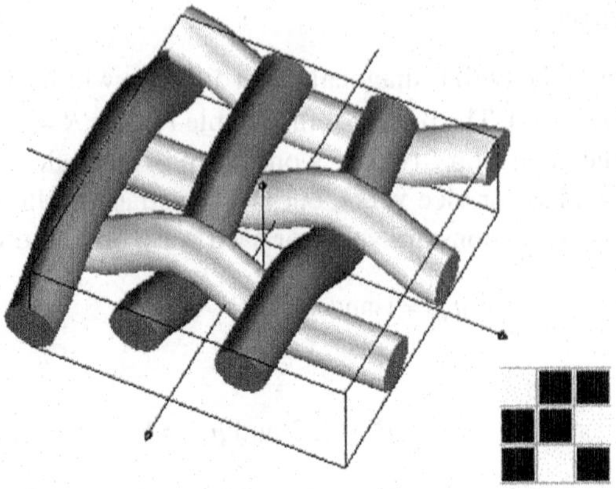

FIGURE 2.7 Twill weave $\frac{2.}{.1}Z$ (warp-faced pattern).

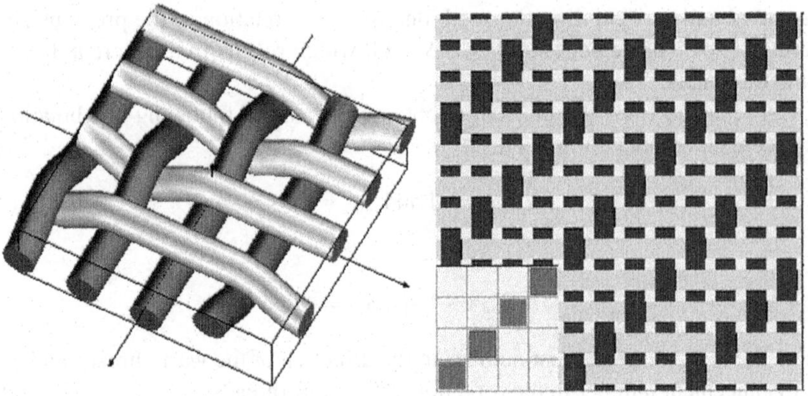

FIGURE 2.8 Right-handed twill weave $\frac{1.}{.3}Z$.

To design a left-handed twill weave (or *S*-twill), the sign of the displacement numbers in warp and weft directions has to be different:

$$S_{wp} = +1 \text{ and } S_{wf} = -1 \tag{2.8}$$

or

$$S_{wp} = -1 \text{ and } S_{wf} = +1 \tag{2.9}$$

The twill diagonals are formed on both sides of the woven fabric; however, if the diagonal is to the left on one side (the technical face), it is to the right on the other side (the technical back) (Figure 2.9).

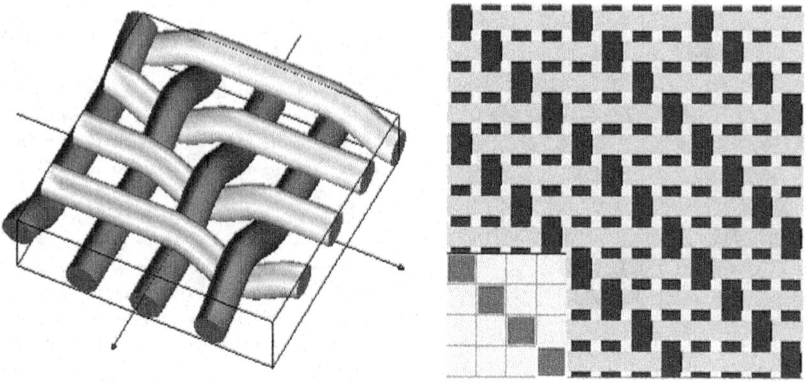

FIGURE 2.9 Left-handed twill weave $\frac{1}{3}S$.

2.2.3 Sateen and Satin Weaves

The *satin* weave is frequently mixed with the *sateen* weave. According to the definition of Grosicki (2004), satin is a warp-faced pattern (Figure 2.10), and sateen is a weft-faced pattern (Figure 2.11). Different warp and weft densities are used to emphasize the warp and weft effects:

- The fabric in the satin weave is produced with higher warp density than weft density to enhance the effects of the warp threads (the ends).
- The fabric in the sateen weave is produced with higher weft density than warp density to bring into relief the weft threads (the picks).

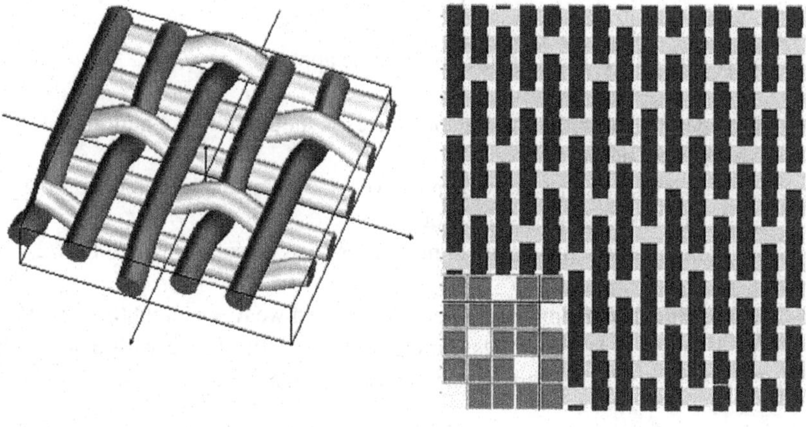

FIGURE 2.10 Satin weave.

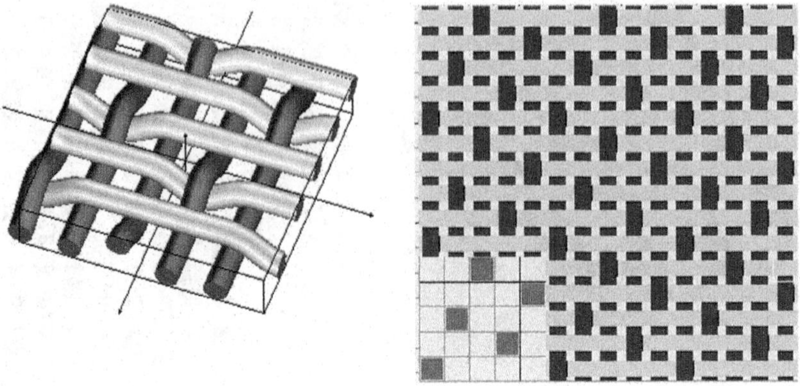

FIGURE 2.11 Sateen weave.

In the repeat of the satin weave, there is one weft overlap alongside a thread and the rest are warp overlaps. The following expressions are valid for each end and pick:

$$n_{wp} = R - 1 \geq 5 \text{ and } n_{wf} = 1 \tag{2.10}$$

In the repeat of the sateen weave, one warp overlap is followed by weft overlaps alongside each thread. These expressions are valid for each end and each pick in the repeat:

$$n_{wp} = 1 \text{ and } n_{wf} = R - 1 \geq 5 \tag{2.11}$$

The displacement number S in satin and sateen weaves can be found, using the expression:

$$1 < S < R - 1 \tag{2.12}$$

At the same time, the following rules have to be applied:

- The repeat of the satin and sateen weaves R should not be a multiple of the displacement number S.
- The repeat of the satin and sateen weaves R and the displacement number S should not have a common divisor.
- It is preferable the displacement number S be as close as possible to $R/2$.

Thus, the possible displacement numbers for a sateen weave on eight threads ($R = 8$) are 3 and 5. Figure 2.12 illustrates the eight-thread satin weave with $S = 3$, while Figure 2.13 shows the eight-thread sateen weave with $S = 5$.

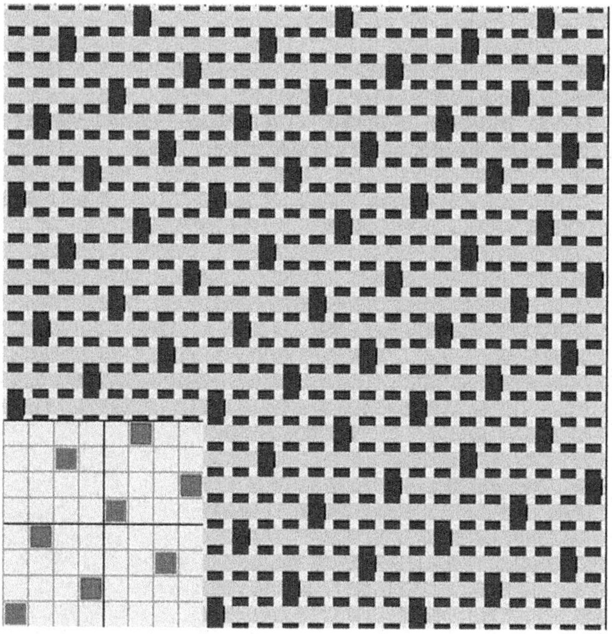

FIGURE 2.12 Satin weave with $R = 8$ and $S = 3$.

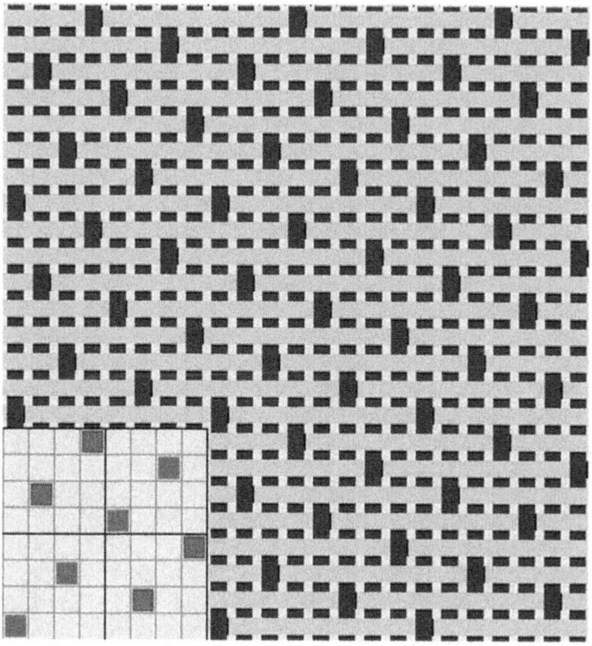

FIGURE 2.13 Sateen weave with $R = 8$ and $S = 5$.

REFERENCES

Damyanov, G., & Chobanov, G. (1988). *Proektirane i stroej na takanite* [Design and Construction of Fabrics], VMEI Lenin, Bulgaria, Sofia.

Grosicki, Z. J. (2004). *Watson's Advanced Textile Design, Compound Woven Structures*, 4th ed. Reprinted by Woodhead, Cambridge, UK.

3 Derivatives of the Elementary Weaves

3.1 BASICS

The derivatives of the elementary weaves are obtained using:

- changes in the way of interlacing between the warp and weft threads, keeping the rules for the construction of the elementary weave;
- changes in the sign and value of the displacement number *S*;
- combination of the two.

3.2 DERIVATIVES OF THE PLAIN WEAVE

Because of the simplicity of the plain weave, it has only two derivative weaves: rib or hopsack.

3.2.1 Rib Weave

The rib is obtained when the basic overlaps of the plain weave are multiplied either on the direction of the warp threads (*warp rib*, Figure 3.1) or on the weft threads (*weft rib*, Figure 3.2). When the number of the warp and weft overlaps are equal, *regular rib* weave is formed (Figures 3.1 and 3.2), while the difference in warp and weft overlaps leads to the construction of an *irregular rib* weave (Figure 3.3).

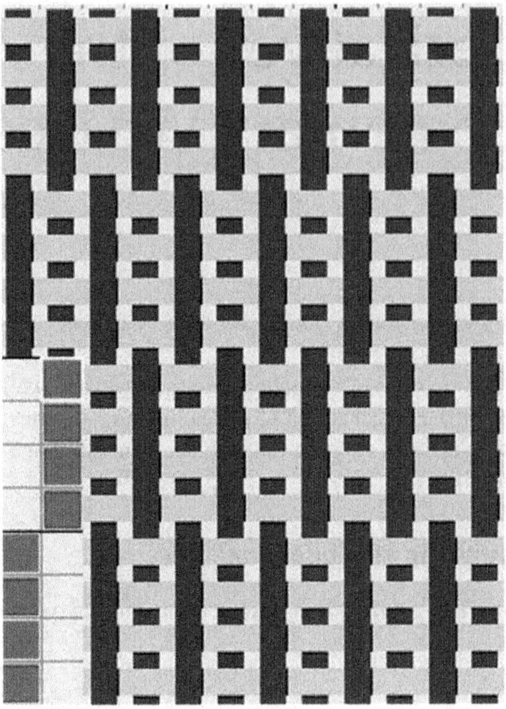

FIGURE 3.1 Regular warp rib $\frac{4}{4}$.

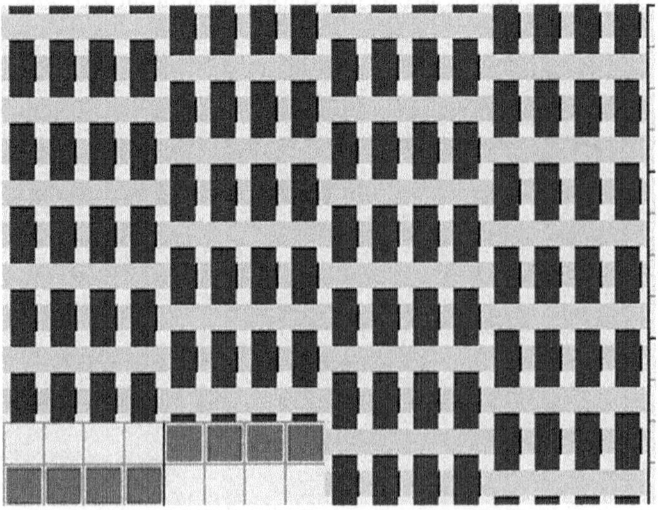

FIGURE 3.2 Regular weft rib $\frac{4}{4}$.

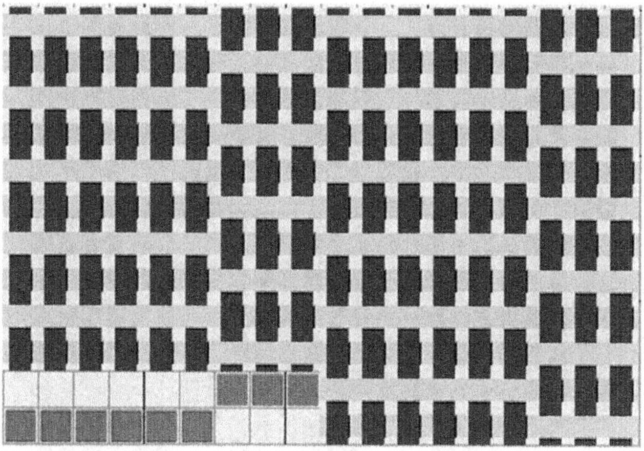

FIGURE 3.3 Irregular warp rib $\frac{5}{3}$.

3.2.2 Hopsack Weave

Hopsack weaves are designed from the plain weave by multiplying the single overlaps in both warp and weft directions. Depending on the number of the additional overlaps in the two directions, the hopsack weave can be *regular* (Figure 3.4) or *irregular* (Figures 3.5 and 3.6). Alongside a single yarn in the repeat of the regular hopsack, it is valid that the warp and weft overlaps are equal, i.e., $n_{wp} = n_{wf}$, while for the irregular hopsack $n_{wp} \neq n_{wf}$.

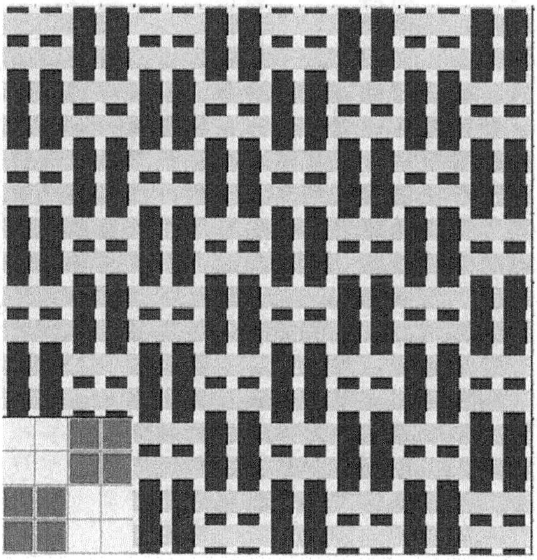

FIGURE 3.4 Regular hopsack weave $\frac{2}{2}$.

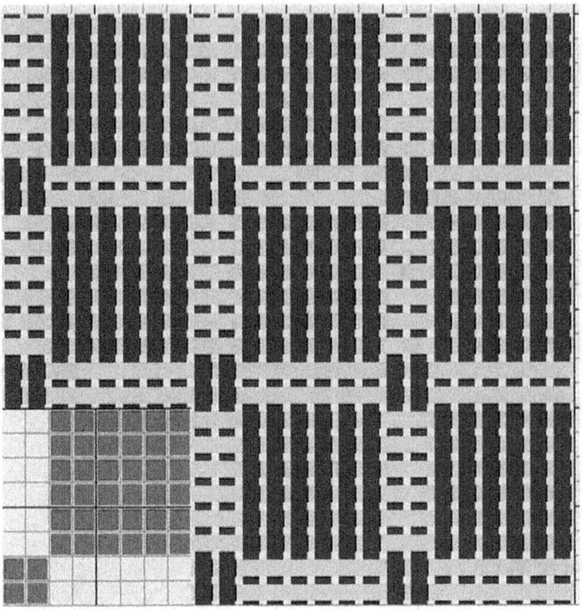

FIGURE 3.5 Irregular hopsack weave $\frac{2}{6}$.

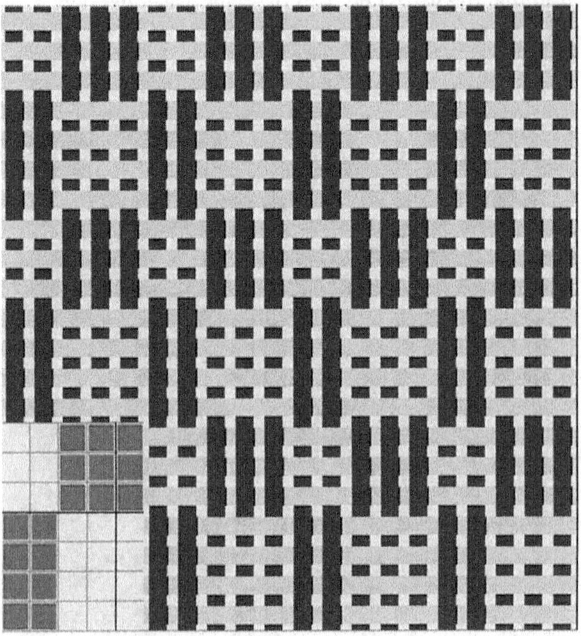

FIGURE 3.6 Irregular hopsack weave.

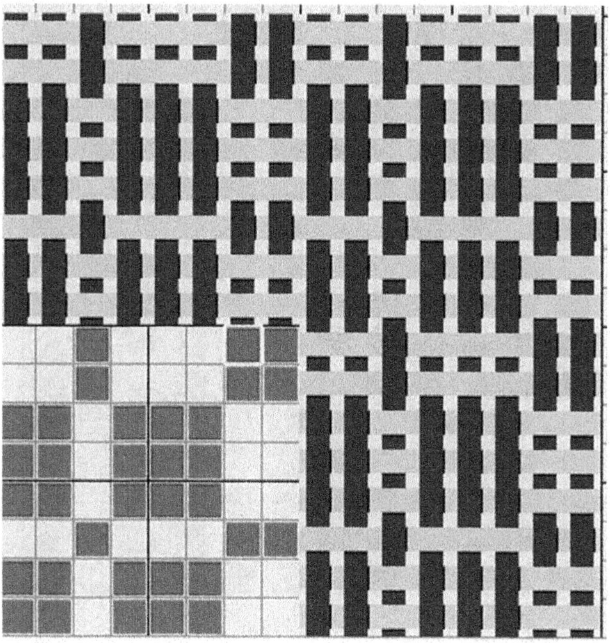

FIGURE 3.7 Fancy hopsack weave.

A version of the hopsack weave is the fancy hopsack, which requires at least two transitions of the weave from warp to weft overlaps. Figure 3.7 presents a version of this weave.

3.3 DERIVATIVES OF THE TWILL WEAVE

3.3.1 Strengthened Twill

The strengthened twill is probably the easiest way to modify the elementary twill weave: the single warp or weft overlap is strengthened by adding one or more overlaps of the same type. If $n_{wp} = n_{wf}$ alongside a single yarn, then diagonals with equal width appear in both sides of the cloth (Figure 3.8). If $n_{wp} \neq n_{wf}$, the warp-faced and weft-faced diagonals have different widths (Figure 3.9).

3.3.2 Complex Twill

The complex twill is obtained when the elementary twill is modified, adding at least two more changes from warp to weft appearance of the twill diagonals. This means that on the first warp thread, the warp overlaps will change to weft overlaps at least twice. Figure 3.10 shows a left-handed twill weave $\frac{3.4}{2.1.}S$.

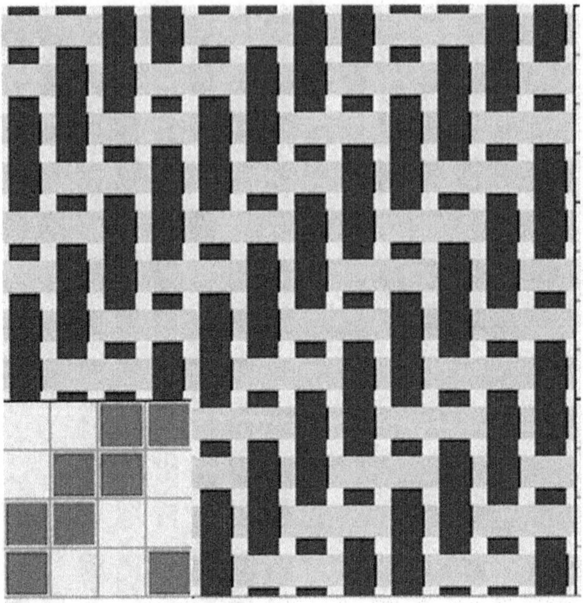

FIGURE 3.8 Strengthened twill $\frac{2}{.2}Z$.

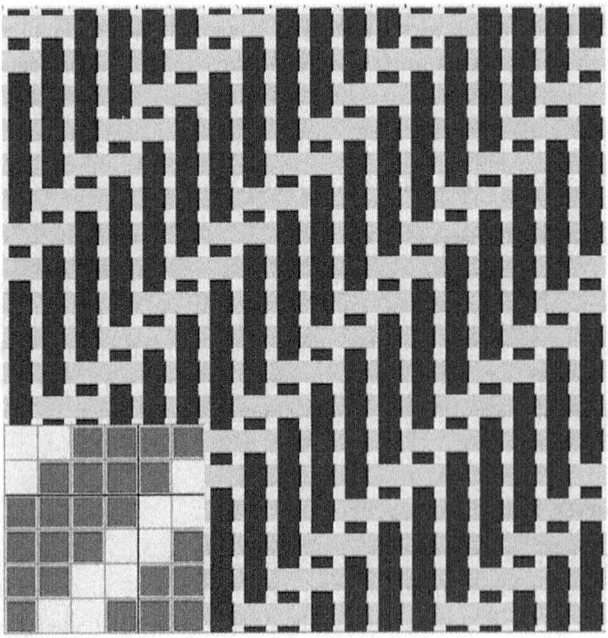

FIGURE 3.9 Strengthened twill weave $\frac{4}{.2}Z$.

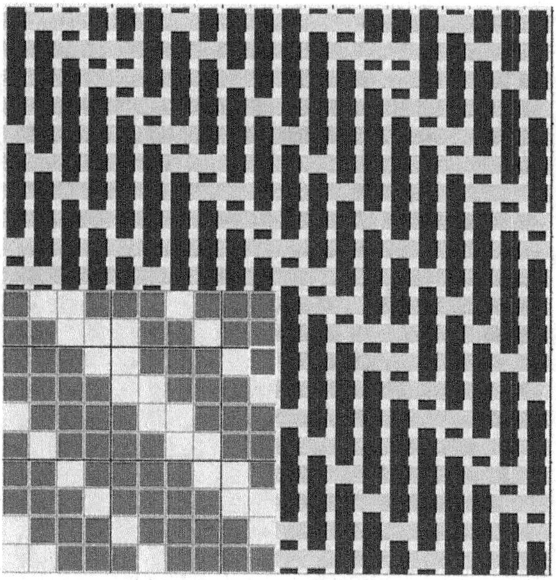

FIGURE 3.10 Complex twill weave $\frac{3.4}{2.1.}S$.

3.3.3 Zig-Zag Twill

The zig-zag or waved twill can be designed based on an elementary twill weave, strengthened twill, or complex twill. After a certain number of threads, the initial weave changes its displacement number S to form left-handed diagonal after the right-handed diagonal of the basic weave or vice versa.

A peak is formed at the point of changing the direction of the diagonal, which may consist of one thread (Figure 3.11) or two threads (Figure 3.12).

The number of threads, after which the direction of the twill diagonal changes, determines the magnitude of the zig-zag effect:

- Reduced (Figure 3.13): when the number of threads W, after which the change of the diagonal occurs, is less than the repeat of the initial weave R, i.e., $W < R$.
- Normal (Figure 3.14): when the number of threads W, after which the change of the diagonal occurs, is equal to the repeat of the initial weave R, i.e., $W = R$.
- Enlarged (Figure 3.15): when the number of threads W, after which the change of the diagonal occurs, is higher than the repeat of the initial weave R, i.e., $W > R$.

Figures 3.11 through 3.15 illustrate zig-zag twill weave designs when the change of the diagonal's direction occurs after several warp threads. The same design principles can be applied to obtain zig-zag twill weaves when the direction of the diagonal

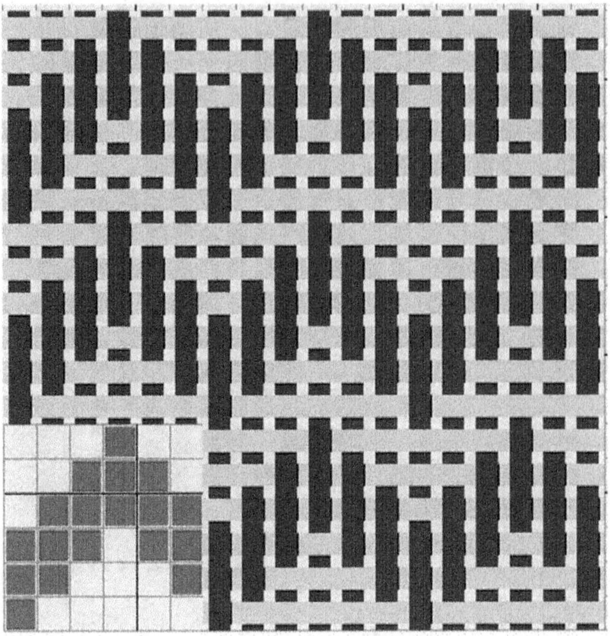

FIGURE 3.11 Zig-zag twill weave with one-thread pick.

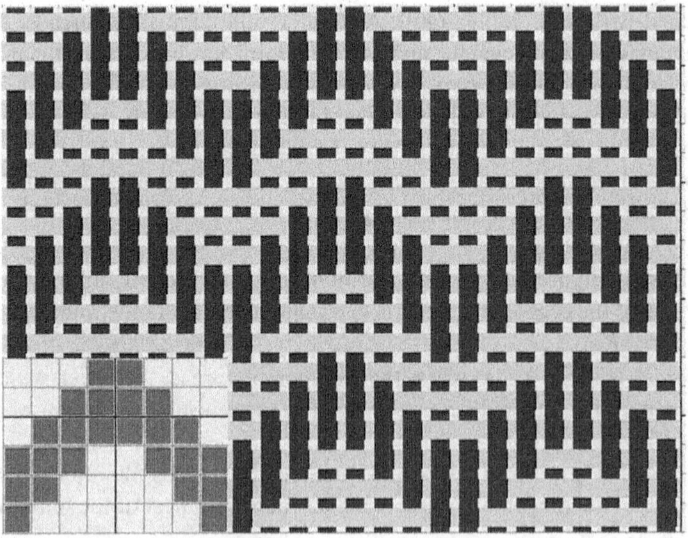

FIGURE 3.12 Zig-zag twill weave with two-thread pick.

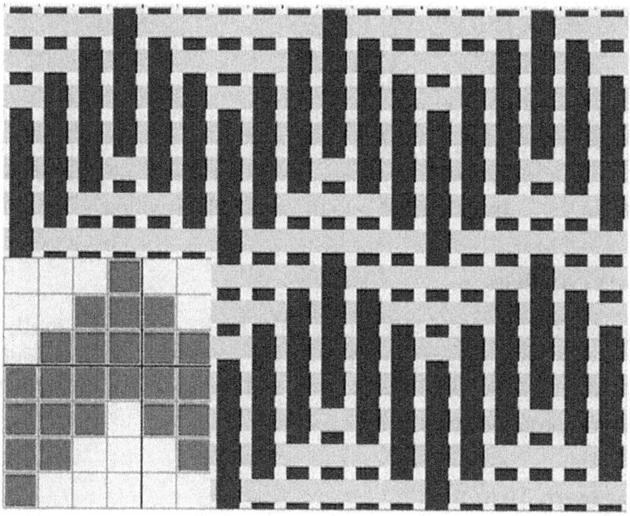

FIGURE 3.13 Zig-zag twill weave with reduced "teeth" ($W = 4$) and one-thread pick from strengthened twill $\frac{4}{.4}Z$.

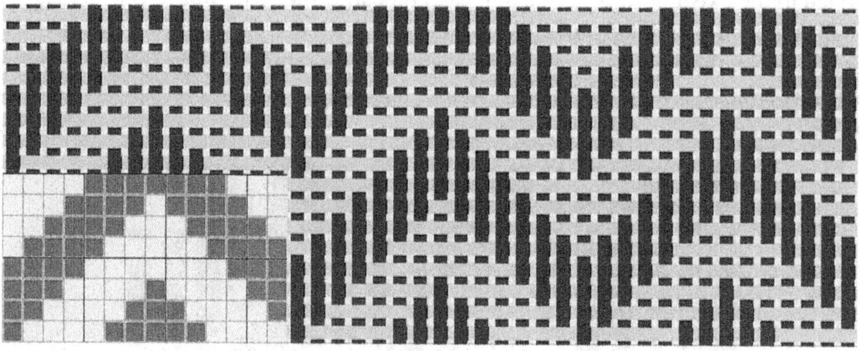

FIGURE 3.14 Zig-zag twill weave with normal "teeth" ($W = R = 8$) and one-thread pick from strengthened twill $\frac{4}{.4}Z$.

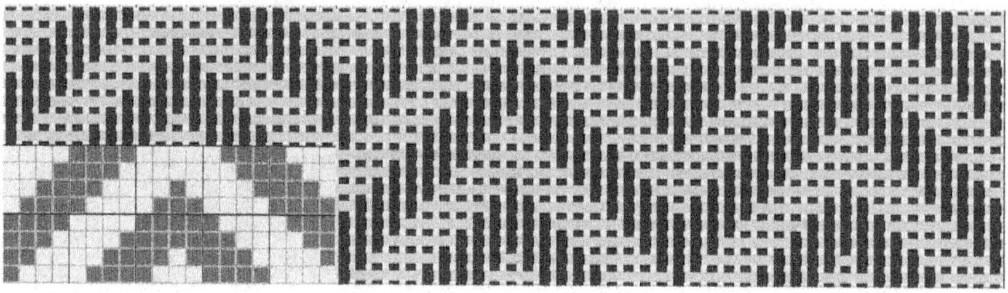

FIGURE 3.15 Zig-zag twill weave with enlarged "teeth" ($W = R = 11$) and one-thread pick from strengthened twill $\frac{4}{.4}Z$.

changes alongside the weft threads. Figure 3.16 shows weft zig-zag twill with two-thread pick and reduced "teeth", while Figure 3.17 illustrates the weft zig-zag twill with one-thread pick and normal "teeth".

If the change in the twill diagonal occurs after a different number of threads in *Z* and *S* directions, the result is a diagonal zig-zag weave. Figure 3.18 shows a diagonal zig-zag twill weave, designed from a complex twill $\frac{2.3.}{.2.1}Z$: four threads of the twill diagonal goes in the *Z* direction and two threads in *S* direction.

3.3.4 Diamond Twill

This weave can be designed using elementary twill, strengthen twill, or complex twill weaves as basic weave patterns. To obtain a diamond twill, the diagonals should simultaneously change in both warp and weft directions. Figure 3.19 shows a diamond twill from a complex twill weave $\frac{1.3.1.}{.2.1.2}Z$.

3.3.5 Shadowed Twill

The design of the shadowed twill starts from an elementary twill weave either with warp or with weft face view. The pattern repeat is built within several repeats of the basic weave. If the basic weave is a warp-faced twill, one additional weft overlap is added to the existing weft overlaps in each repeat; if the basic weave is a weft-faced twill, one additional warp overlap is added to the existing warp overlaps in each repeat. Thus, the basic weave converts to its negative representation. Then, the additional overlaps decrease to the basic weave, plus one additional overlap.

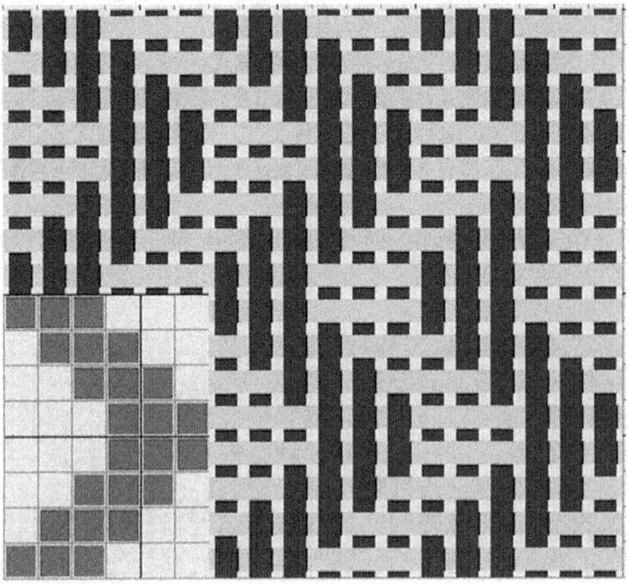

FIGURE 3.16 Zig-zag weft twill weave with reduced "teeth" $(W = 4)$ and two-thread pick from strengthened twill $\frac{3}{3}Z$.

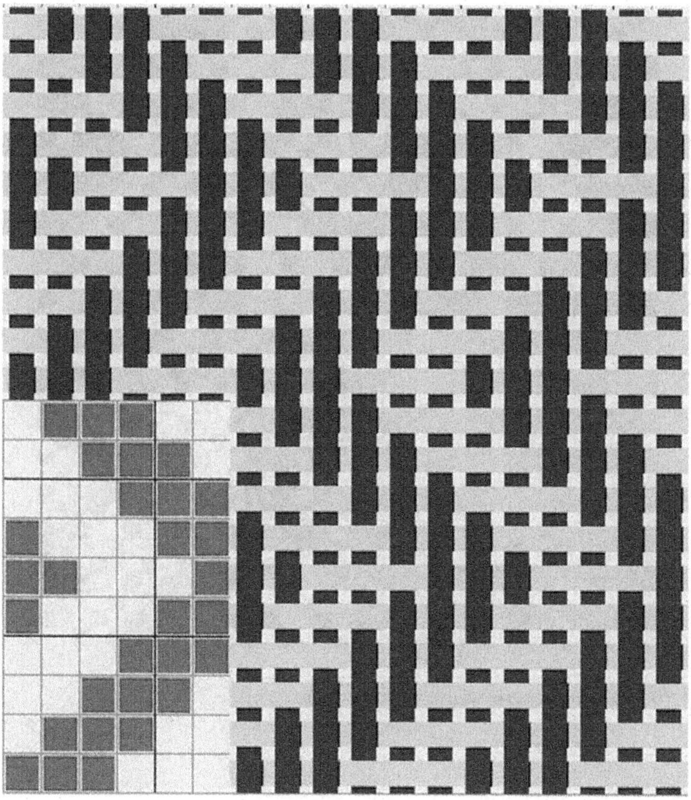

FIGURE 3.17 Zig-zag twill weave with normal "teeth" $(W = R = 6)$ and one-thread pick from strengthened twill $\frac{3.}{.3}Z$.

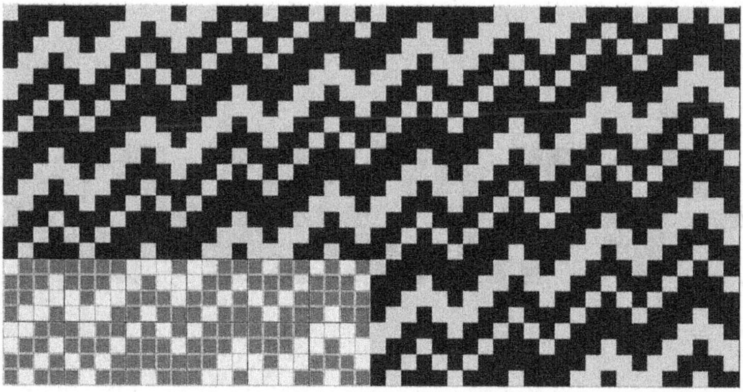

FIGURE 3.18 Zig-zag diagonal twill weave from complex twill $\frac{2.3.}{.2.1}Z$.

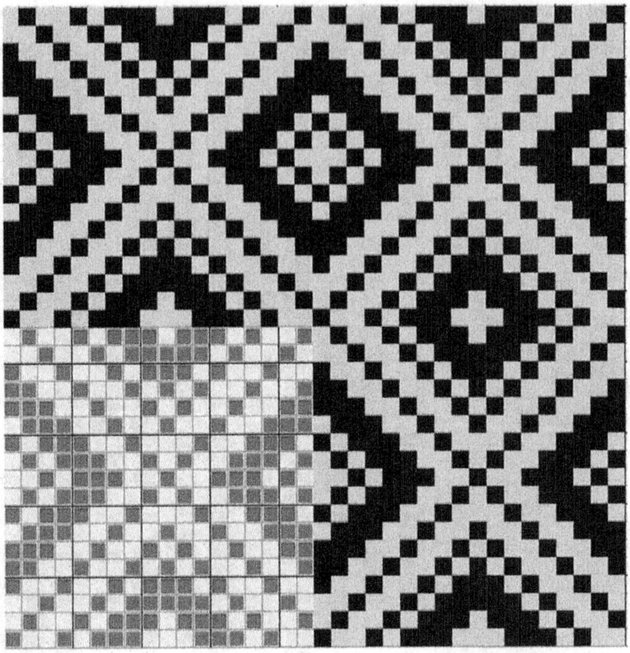

FIGURE 3.19 Diamond twill weave from complex twill $\frac{1.3.1.}{.2.1.2}Z$.

Figure 3.20 presents the repeat of a shadowed twill from a basic twill $\frac{1.}{.4}Z$. The different repeats of the basic twill are separated, and the number of additional warp overlaps is marked. The warp repeat of the new weave is $R_{wp} = 30$, and the weft repeat is equal to the repeat of the basic weave $R_{wf} = 5$.

For a better impression of the shadowed effect, three repeats on the warp direction and ten repeats on the direction of the weft are presented in Figure 3.21a. The cloth texture is shown in Figures 3.21b and 3.22.

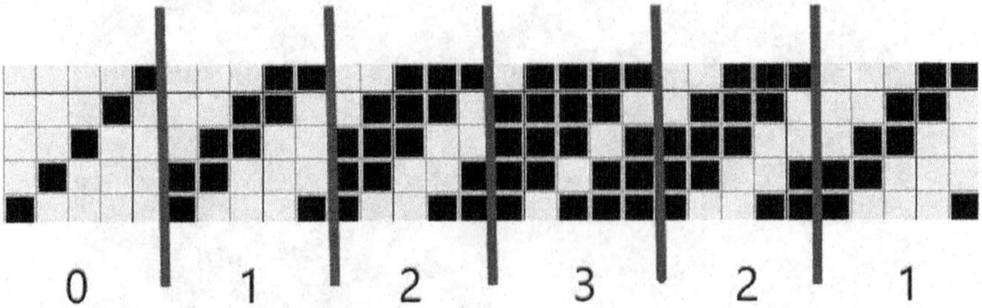

FIGURE 3.20 Shadowed twill weave from a basic twill $\frac{1.}{.4}Z$.

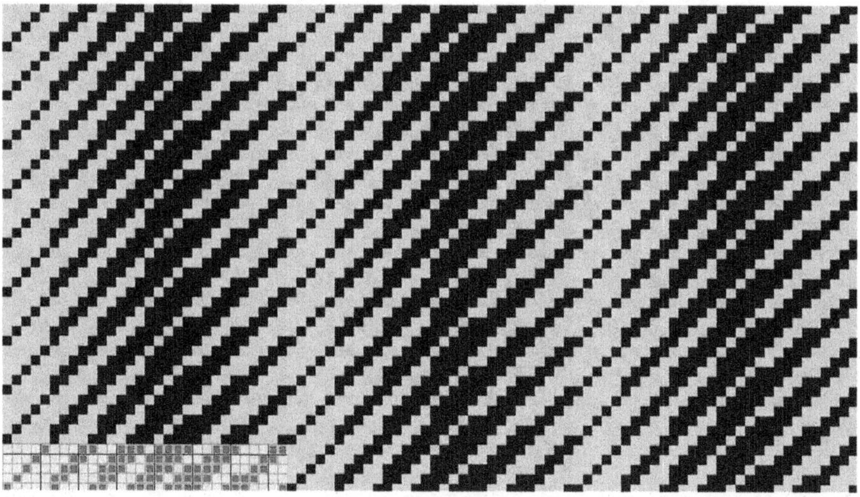

(a)

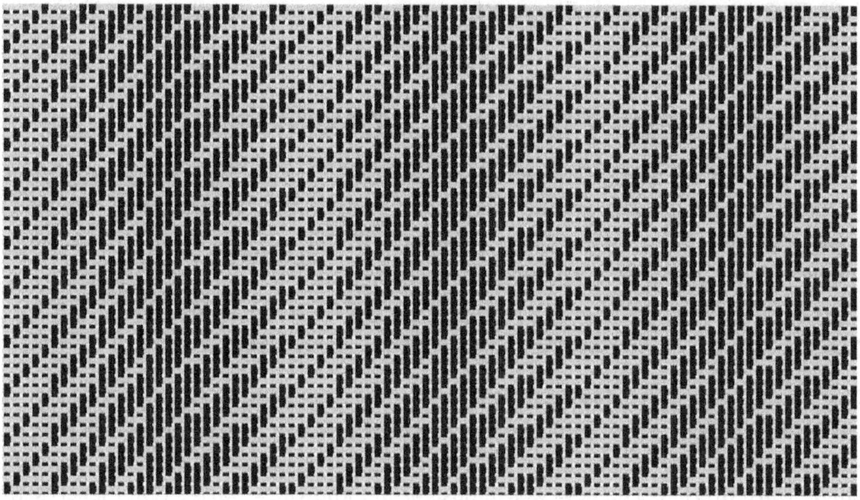

(b)

FIGURE 3.21 Shadowed twill weave from a basic twill $\frac{1}{4}Z$: (a) colored view of black warp and white weft threads and (b) fabric view.

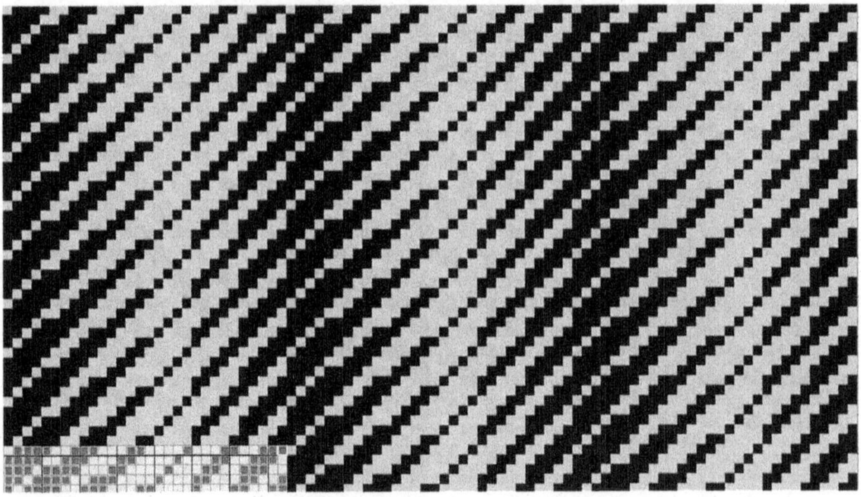

FIGURE 3.22 Shadowed twill weave from a basic twill $\frac{4}{.1}Z$: colored view of black warp and white weft threads.

3.4 DERIVATIVES OF SATIN AND SATEEN WEAVES

The derivatives of satin and sateen weaves are based on the change of the single weft overlap (for a satin weave) or the single warp overlap (for a sateen weave) alongside a single thread in the weave repeat. Additional overlaps of the same type are added and used to decrease the yarn float.

3.4.1 Strengthened Satin or Strengthened Sateen Weaves

The new design adds additional weft overlaps in the satin weave, placed under or above the existing single weft overlap. If a sateen weave is used as a basic pattern, then additional one or more warp overlaps are added under or above the existing single warp overlap. The new weave designs are known as warp-strengthened satin or warp-strengthened sateen weaves.

Figure 3.23 presents warp-strengthened satin weave $\frac{1}{.6}(S = 3)$ with one additional weft overlap (in the direction of the warp threads), while Figure 3.24 shows a warp-strengthened sateen weave $\frac{1}{.6}(S = 3)$ with three additional weft overlaps (in the direction of the warp threads).

The same design procedure can be applied on the direction of the weft threads, adding one or more new weft overlaps on the left or right side of the existing in the basic satin weave single weft overlap. If the basic weave is sateen, then warp overlaps are added.

Figure 3.25 illustrates a weft-strengthened satin weave $\frac{7}{.1}(S = 5)$ with one additional weft overlap. Figure 3.26 shows a weft-strengthened sateen weave $\frac{1}{.7}(S = 3)$ with two additional weft overlaps.

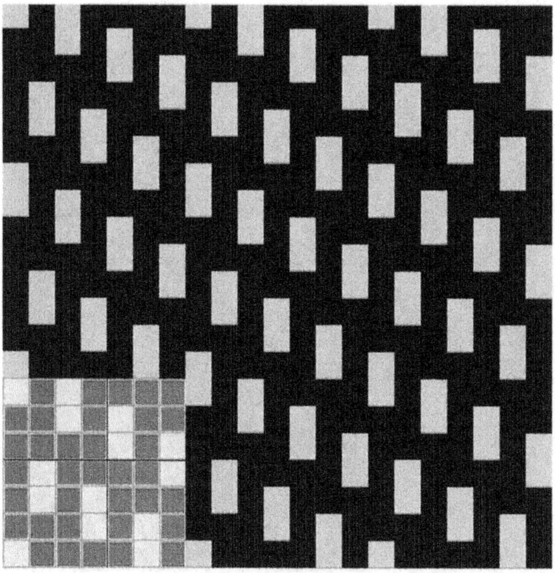

FIGURE 3.23 Warp-strengthened satin weave from satin $\frac{6}{.1}$ $(S = 3)$ with one additional weft overlaps (in the direction of the warp threads).

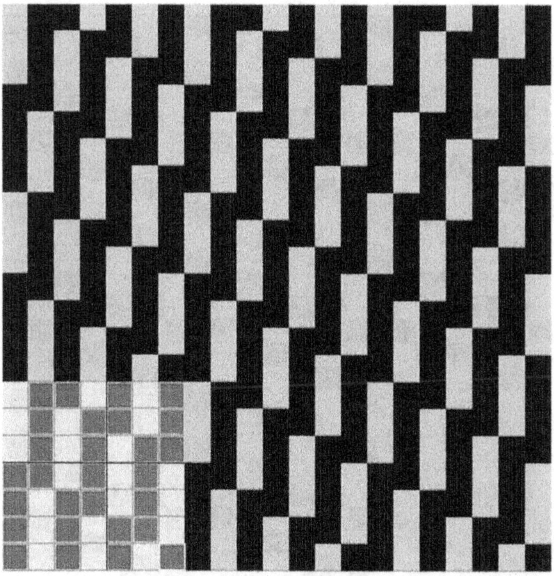

FIGURE 3.24 Warp-strengthened sateen weave from sateen $\frac{1}{.6}$ $(S = 3)$ with three additional warp overlaps (in the direction of the warp threads).

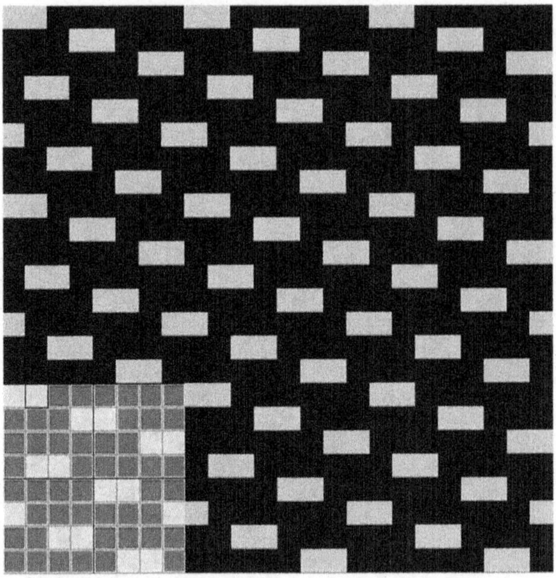

FIGURE 3.25 Weft-strengthened satin weave from sateen $\frac{7}{.1}(S=5)$ with one additional weft overlap (in the direction of the weft threads).

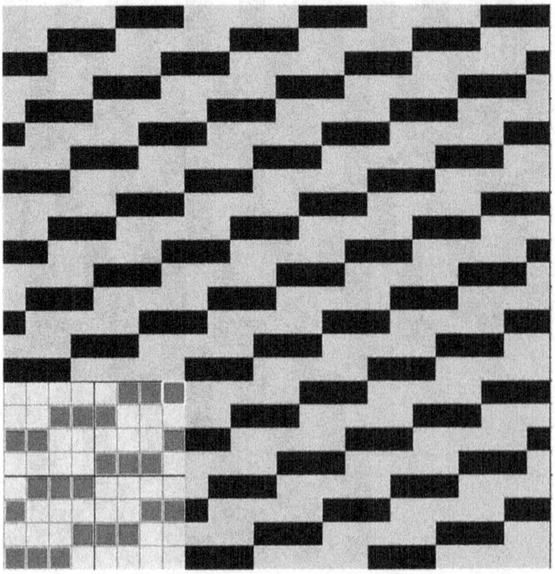

FIGURE 3.26 Weft-strengthened sateen weave from sateen $\frac{1}{.7}(S=3)$ with two additional warp overlaps (in the direction of the weft threads).

3.4.2 Shadowed Sateen Weave

The design of this weave pattern usually starts from a sateen weave. Several repeats of the basic weave are used, and, in each repeat, one additional warp overlap is added to the already existing warp overlaps. Thus, the basic sateen weave converts to its negative representation: a satin weave. Then, the additional overlaps decrease to the basic weave, plus one additional overlap.

The method is quite similar to the design of both a shadowed twill and a strengthened sateen weave. Usually, the additional warp overlaps are added in the direction of the warp threads, as shown in Figure 3.27. However, a very similar design can be obtained by adding the additional warp overlaps on the left or right side of the existing single overlap, following the weft threads (Figure 3.28).

The weave patterns, shown in Figures 3.27 and 3.28, are designed using five-thread sateen weave $\frac{1}{.4}(S=2)$ as a basic pattern. Three repeats on the warp and three repeats on the weft directions are illustrated.

In fact, the same cloth design can be obtained, starting from a satin weave and marking additional weft overlaps to the basic weave.

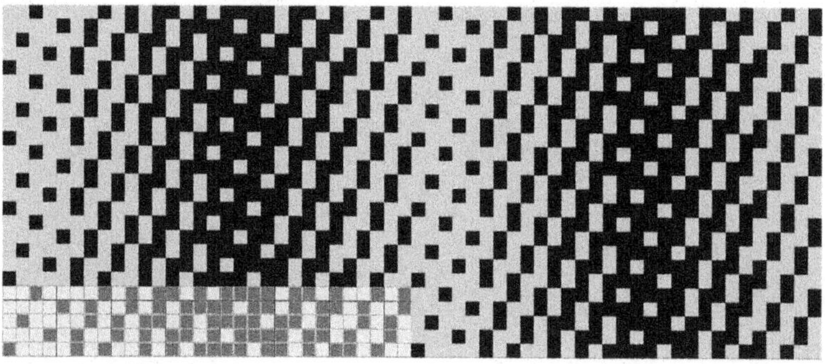

FIGURE 3.27 Shadowed sateen weave from sateen $\frac{1}{.4}(S=2)$ with additional overlaps on the direction of the warp threads.

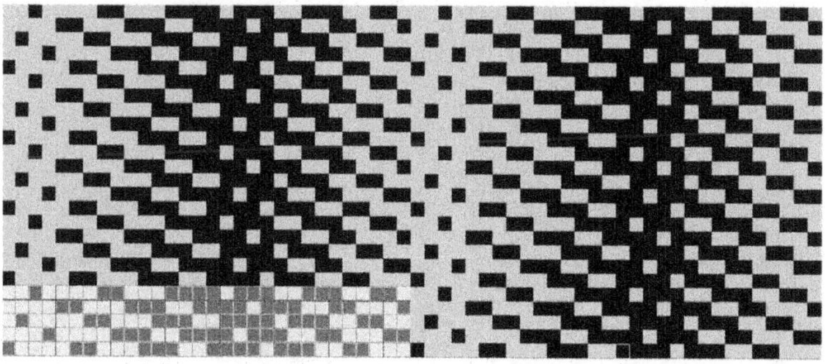

FIGURE 3.28 Shadowed sateen weave from sateen $\frac{1}{.4}(S=2)$ with additional overlaps on the direction of the weft threads.

3.4.2 Shadowed Sateen Weave

The design of this weave pattern usually starts from a sateen weave. Several repeats of the basic weave are used and, in each repeat, one additional warp overlap is added to the already existing warp overlaps. Thus, the basic sateen weave would be its negative representation, a sateen weave. [illegible] the additional [illegible] base weave, adds one additional overlap.

The method is quite similar to the design of both a shadow twill and sateen weaves. Here the additional warp overlaps are added in the direction of the warp thread, as shown in Figure 4.27. However, a very similar design can be obtained by adding the additional overlaps [illegible] to the left or the right of the existing single overlaps in the weft thread (Figure 4.28).

The weave patterns shown in Figures 4.27 and 4.28 [illegible] thread sateen weave as a basic pattern. Three repeats [illegible] and three repeats in the weft direction are illustrated.

In fact, this sateen cloth design can be considered as [illegible] from a [illegible] weave and [illegible] adding additional weft overlaps to the base weave.

FIGURE 4.27 Shadowed sateen weave from [illegible] with additional overlaps in the direction of the warp threads.

FIGURE 4.28 Shadowed sateen weave from sateen [illegible] with additional overlaps in the direction of the weft threads.

4 New Weave Patterns Based on Patterns Merge

4.1 WEAVE PATTERNS FOR STRIPE EFFECTS

The visual stripe effect of this group of weave patterns appears due to different light refraction of the compound (basic) patterns. The weaving effect might be enhanced if threads of different colors are used for each of the patterns.

4.1.1 Vertical Stripes of Miscellaneous Weave Patterns

The resulting new weave pattern is a combination of elementary weaves, derivatives of the elementary weaves, or more complex weave patterns. The basic weaves should have clear texture and small repeats (*R*). The first warp thread of the next weave is good to interlace negatively to the last thread of the previous weave (the warp overlaps to change to weft overlaps and vice versa) to obtain distinct contours between the stripes. If the combination of the basic weaves does not allow a negative transition between last and first threads of each stripe, one or more split warp threads may be placed between the two adjacent stripes.

The number of the compound weaves in the repeat of the new weave usually varies from 2 to 6 and is limited by:

- the warp repeat of the compound weaves R_{wp};
- the number of harnesses on the loom.

The repeat of the new weave in warp direction R_{wp} can be calculated as:

$$R_{wp} = \sum\nolimits_{i=1}^{k} n_{wpi} \tag{4.1}$$

where n_{wpi} is the number of the warp threads in each compound weave and k is the number of the compound weaves.

The weft repeat R_{wt} of the new weave is the least common multiple (LCM) of the weft repeat of the compound weaves R_{wfi}, i.e.,

$$R_{wf} = \text{LCM}\left(R_{wfi}\right) \tag{4.2}$$

Figure 4.1 is a combination between two repeats of a hopsack weave $\frac{2.}{.2}$ and two repeats of a twill weave $\frac{2.}{.2}Z$. There is a negative transition between the last warp

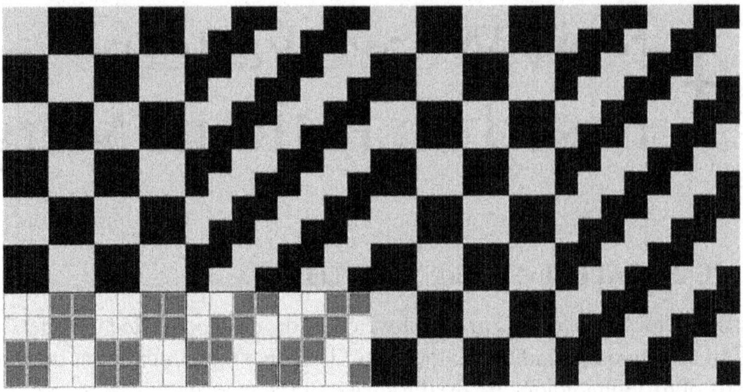

FIGURE 4.1 Vertical stripes from two repeats of a hopsack weave $\frac{2}{.2}$ and two repeats of a twill weave $\frac{2}{.2}Z$.

thread of the hopsack weave and the first thread of the twill weave; however, the negative transition between the last warp thread of the twill and the first thread of the next stripe in a hopsack weave is not possible. A solution is to merge a third weave, e.g., plain weave on two threads, to assure a clearer contour of the stripe.

Figure 4.2 shows a weave for vertical stripes, based on the merge of two strengthened twill weaves $\frac{3}{.3}$ with Z and S diagonals. The second weave starts in a way to obtain a negative transition between the compound weaves. This type of weave is also known as a herringbone twill.

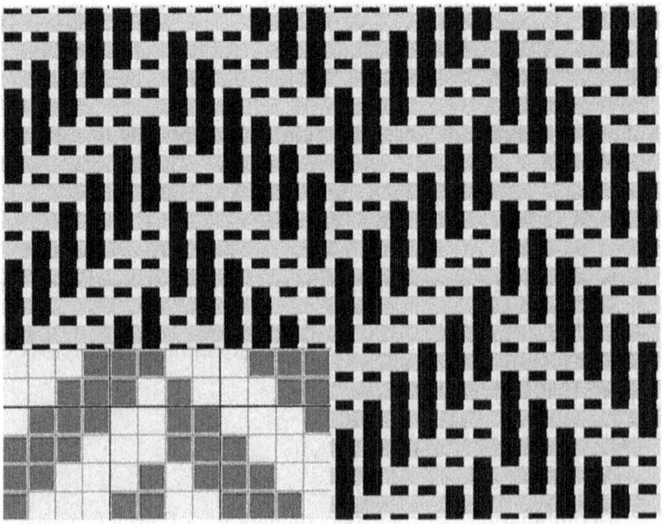

FIGURE 4.2 Vertical stripes from two repeats of a hopsack weave $\frac{2}{.2}$ and two repeats of a twill weave $\frac{2}{.2}Z$.

4.1.2 Horizontal Stripes of Miscellaneous Weave Patterns

It is possible to merge more than six weave patterns to obtain the effect of horizontal stripes on the cloth. The reason is that there are no limitations, concerning the number of the harnesses on the loom and the warp repeat of the final weave.

The warp repeat R_{wp} of the resulting weave can be calculated as the LCM of the warp repeats of the compound weaves R_{wri}, i.e.,

$$R_{wp} = \text{LCM}\left(R_{wpi}\right) \quad (4.3)$$

The weft repeat R_{wt} of the resulting weave is a sum of the weft threads in each compound weave:

$$R_{wf} = \sum\nolimits_{i=1}^{n} n_{wfi} \quad (4.4)$$

where n_{wfi} is the number of the weft threads in each compound weave and i is the number of the compound weaves.

Figure 4.3 shows the effect "horizontal stripes" on the cloth after merging satin and sateen weaves ($R = 8$, $S = 3$).

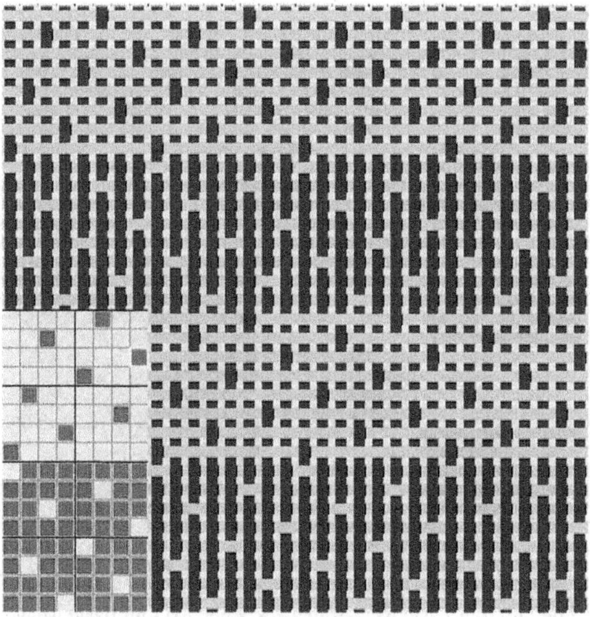

FIGURE 4.3 Horizontal stripes from satin and sateen weaves ($R = 8$, $S = 3$).

4.2 WEAVE PATTERNS FOR CELL-LIKE EFFECTS

The cell-like weaving effect is due to the different light reflection of the interlacing of the threads in the compound weaves. The way of merging of the compound weaves leads to cell-like cloth effects with a square or rectangular shape.

4.2.1 Square Weaves

The weaves are inserted in square cells, usually in a 2 × 2 net. However, 3 × 3 net and even bigger can also be used. The result is that not only square but rectangular cells may also appear on the fabric surface.

Figure 4.4 shows different possibilities in the designing (each color is considered as a different weave pattern):

- Two basic weaves are staggered in the repeat of the resultant weave (Figure 4.4a).
- Three basic weaves are placed in the four quadrants, so one of them is staggered in the repeat (Figure 4.4b).
- Four basic weaves are merged in each of the four quadrants (Figure 4.4c).
- The cell-like effect can be obtained merging the basic weaves in a 3 × 3 net. Figure 4.4d shows the possible disposition of four basic weaves in nine cells.
- The proper grouping of three patterns in a 3 × 3 net leads to the appearance of squares with different size and rectangles (Figure 4.4e).

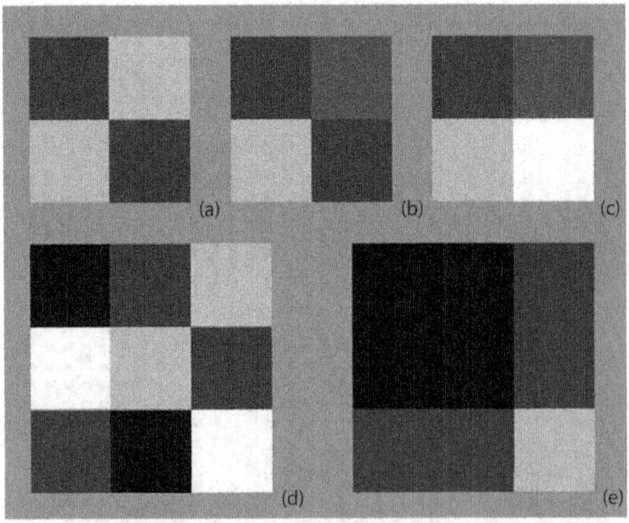

FIGURE 4.4 Cell-like weave effects: (a) two basic weaves in a 2 × 2 net; (b) three basic weaves in a 2 × 2 net; (c) four basic weaves in a 2 × 2 net; (d) four basic weaves in a 3 × 3 net; and (e) three basic weaves in a 3 × 3 net.

The repeats of the basic weaves must be the same or multiples of one another numbers. The warp repeat of the cell-like weave is:

$$R_{wp} = \sum_{i=1}^{k} n_{wpi} \tag{4.5}$$

where n_{wpi} is the number of the warp threads in each compound weave and k is the number of the compound weaves.

The weft repeat of the cell-like weave is:

$$R_{wf} = \sum_{i=1}^{k} n_{wfi} \tag{4.6}$$

where n_{wfi} is the number of the weft threads in each compound weave.

If the cell-like weave consists of 2 × 2 cells, the warp and weft repeats can be calculated as:

$$R_{wp} = 2n_{wp} \tag{4.7}$$

$$R_{wf} = 2n_{wf} \tag{4.8}$$

where n_{wp} and n_{wf} are the number of warp and weft threads in one cell, respectively.

The transition between the two neighboring cells should be negative to obtain a sharp change in the weave effect. The yarn crimp in the basic weave has to be similar to avoid differences in the strain of the warp threads wound onto one warp-beam.

Figure 4.5 shows a square weave based on 2 × 2 net with a plain weave and a hopsack weave $\frac{2\,.}{.\,2}$. Each cell consists of eight yarns.

Figure 4.6 illustrates how square weaves can be used to obtain new design effects. A 3 × 3 net is used, and three basic weaves: a plain weave, twill weave $\frac{2\,.}{.\,2}Z$, and a twill weave $\frac{2\,.}{.\,2}S$. A negative transition between the cells with twill weaves is assured. The resulting design is shown in Figure 4.7.

4.2.2 Fancy Basket Weaves

The fancy basket weave combines a plain weave, rib, and hopsack weave in a square repeat. The selected combinations of warp and weft overlaps are presented in the same way as for the complex twill weave: at least two changes in the yarn interlacing from warp to the weft and vice versa are required. The presence of a single overlap (warp or weft) is also needed.

Figure 4.8 presents a fancy basket weave $\frac{2\,.\,3\,.\,1\,.}{.\,4\,.\,1\,.\,2}$, and Figure 4.9 shows a fancy basket weave $\frac{.\,1\,.\,2\,.\,4}{3\,.\,1\,.\,3\,.}$.

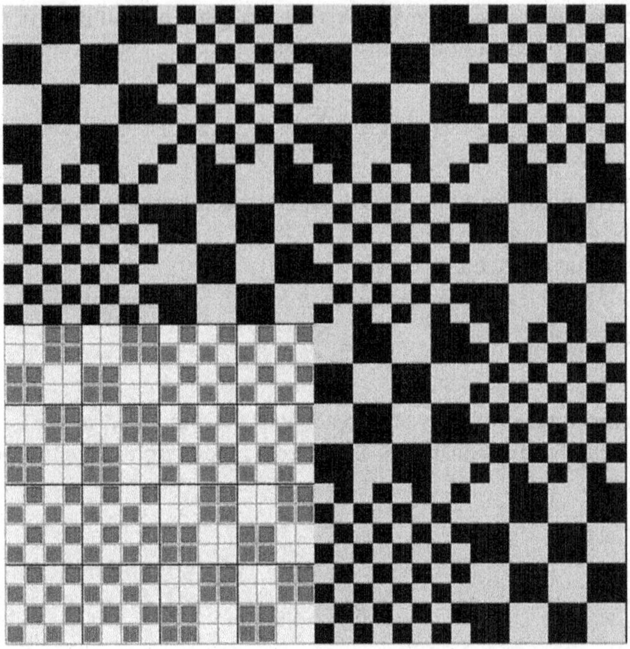

FIGURE 4.5 Square weave from a plain weave and a hopsack weave $\frac{2.}{.2}$ (2 × 2 net, 8 yarns in each cell).

twill $\frac{2\,.}{.\,2}S$	plain weave	plain weave
twill $\frac{2\,.}{.\,2}S$	plain weave	plain weave
twill $\frac{2\,.}{.\,2}Z$	twill $\frac{2\,.}{.\,2}S$	twill $\frac{2\,.}{.\,2}S$

FIGURE 4.6 Square weave from three basic weaves in a 3 × 3 net.

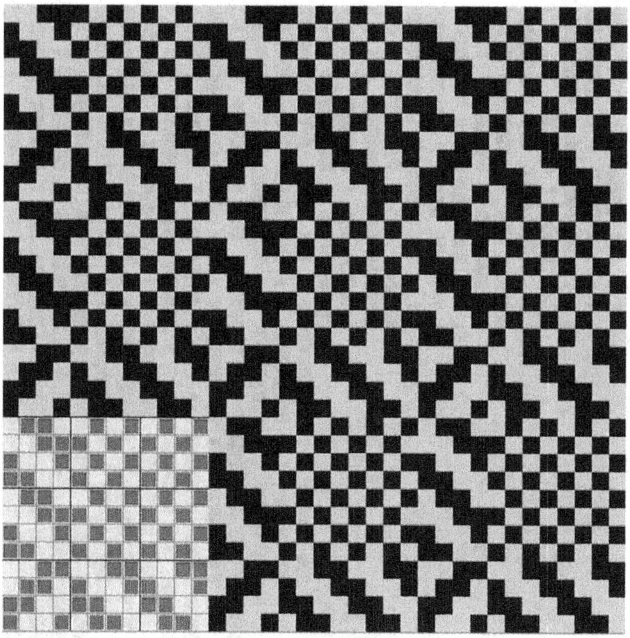

FIGURE 4.7 Resulting weave from the merging plan in Figure 4.6.

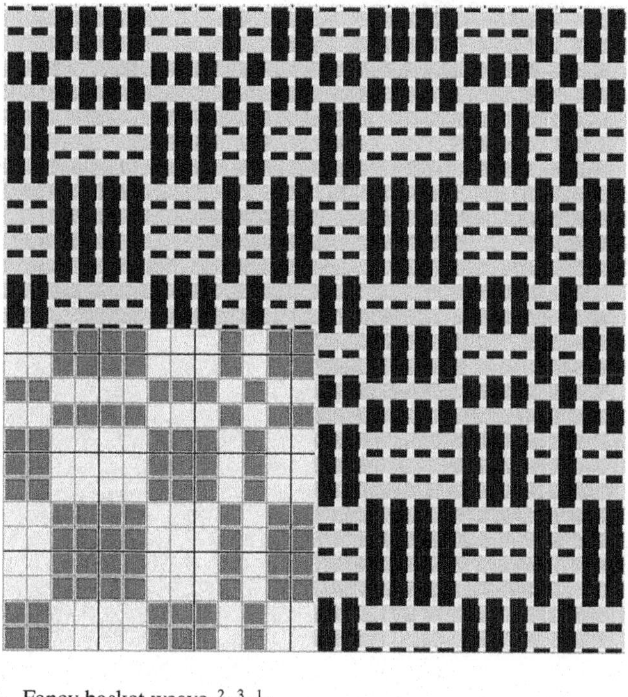

FIGURE 4.8 Fancy basket weave $\frac{2\,.\,3\,.\,1\,.}{.\,4\,.\,1\,.\,2}$.

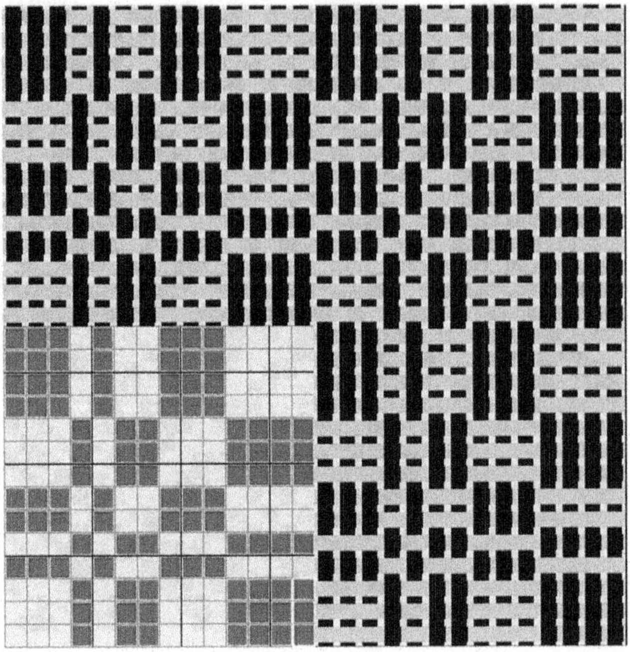

FIGURE 4.9 Fancy basket weave $\frac{1\ .\ 2\ .\ 4}{3\ .\ 1\ .\ 3\ .}$.

5 New Weave Patterns Based on Motifs

5.1 WEAVE PATTERNS FOR CELL-LIKE EFFECTS

The weaving effects of the patterns for cell-like effects are different orthogonal figures.

The new weave is based on a motif, which determines the type and the position of the geometric figures. The motif may be of different size and shape and can be:

- a weave pattern;
- a composition that follows the rules of a known weave pattern;
- a free composition.

Two types of basic weaves are used to fill in the cells of the motif:

- a weave, which fills the motif (its full cells);
- a weave, which fills the background (the empty cells).

The motif is usually augmented so as each of its cells include several warp and weft threads; thus, more than one repeat of the selected basic weaves takes up one cell.

5.1.1 Weaves Based on Basket Motifs

The main motif in this weave pattern is the fancy basket weave. The basic weaves, used to fill in the cells of the motif, are of warp face and weft face with a repeat $R \leq 5$, e.g., warp-faced and weft-faced elementary twill, satin and sateen, etc.

The repeats of the basic weaves must be the same or multiples of one another numbers. The warp face weave is used to fill the cells of the motif and the weft face weave fills the background cells. The best results are obtained when the last thread from the weave for the motif interlace negatively to the first thread of the weave for the background and vice versa. Usually, the motif is not augmented, i.e., each cell of the motif is filled with one repeat of the basic weave only.

The warp repeat R_{wp} of the weave is calculated as:

$$R_{wp} = R_{wp,m}.R_{wp,b} \tag{5.1}$$

where $R_{wp,m}$ is the warp repeat of the motif and $R_{wp,b}$ is the warp repeat of the basic weaves (assuming that the repeats of the basic weaves are the same).

The weft repeat R_{wf} of the weave is calculated similarly:

$$R_{wf} = R_{wf,m}.R_{wf,b} \quad (5.2)$$

where $R_{wf,m}$ is the weft repeat of the motif and $R_{wf,b}$ is the weft repeat of the basic weaves (assuming that the repeats of the basic weaves are the same).

Figure 5.1a presents a fancy basket weave $\frac{3.1.}{.1.2}$, used as a motif for the design of a new, cell-like weave pattern. The basic patterns are twill $\frac{3.}{.1}Z$ (for the cells of the motif, Figure 5.1b) and twill $\frac{1.}{.3}S$ (for the cells of the background, Figure 5.1c). The result of the design is shown in Figure 5.2. The new weave pattern looks complex enough but requires only eight harnesses. Figure 5.3 presents the cloth texture.

5.1.2 Figural Rib

The figural rib weave is designed using a motif and two basic weaves: a warp rib for the cells of the motif and a weft rib weave for the cells of the background. The long repeats of the basic weaves must be the same or multiples of one another numbers.

The cells of the motif are augmented 2, 3, or more times the longer repeat of the rib weave, i.e., 2, 3, or more (longer) repeats of the rib weave are used to fill one cell of the motif.

The warp repeat R_{wp} of the new weave pattern can be calculated from:

$$R_{wp} = A.R_{wp,m}.R_{wp,b} \quad (5.3)$$

where $A = 1,2,...,n$ is the augmentation of each cell of the motif and $R_{wp,b}$ is the warp repeat of the warp rib.

The weft repeat R_{wf} of the weave is determined by analogy:

$$R_{wf} = A.R_{wf,m}.R_{wf,b} \quad (5.4)$$

where $R_{wf,b}$ is the weft repeat of the warp rib.

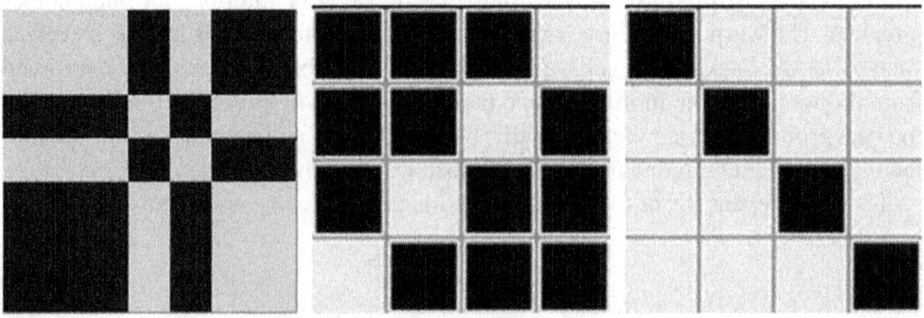

(a) (b) (c)

FIGURE 5.1 Cell-like weave based on a basket weave motif: (a) motif from a fancy basket weave $\frac{3.1.}{.1.2}$; (b) twill $\frac{3.}{.1}Z$ as a basic pattern for the cells of the motif; and (c) twill $\frac{1.}{.3}S$ as a basic pattern for the cells of the background.

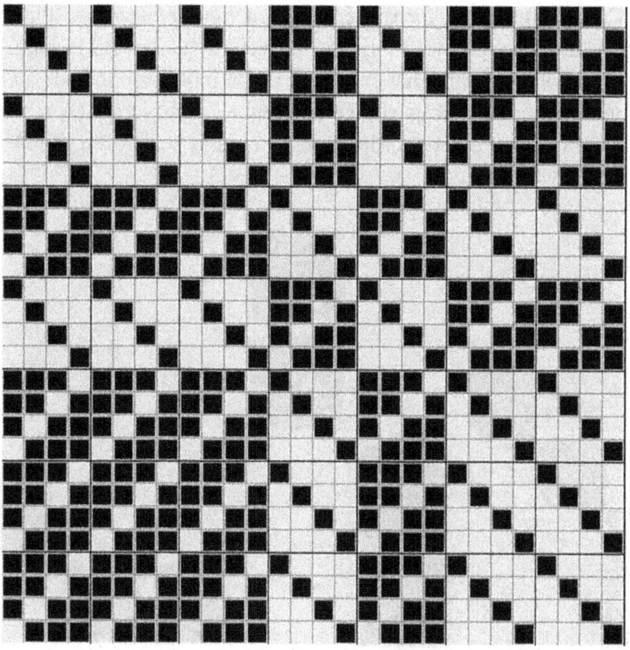

FIGURE 5.2 Cell-like weave based on a basket weave motif: the repeat.

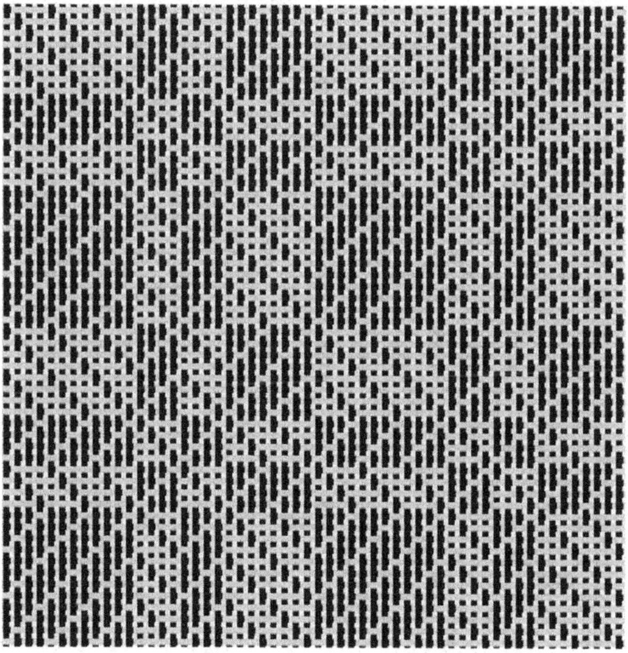

FIGURE 5.3 Cell-like weave based on a basket weave motif: the cloth texture.

Figure 5.4 shows the motif for the design of a figural rib weave together with the selected warp rib for the cells of the motif and the weft rib for the background. The selected augmentation is 2, which means that every cell of the motif will include an 8 × 8 net.

The repeat of the figural rib is shown in Figure 5.5. Figure 5.6 presents the color design of the fabric (three repeats in both warp and weft directions): black warp threads and white weft threads.

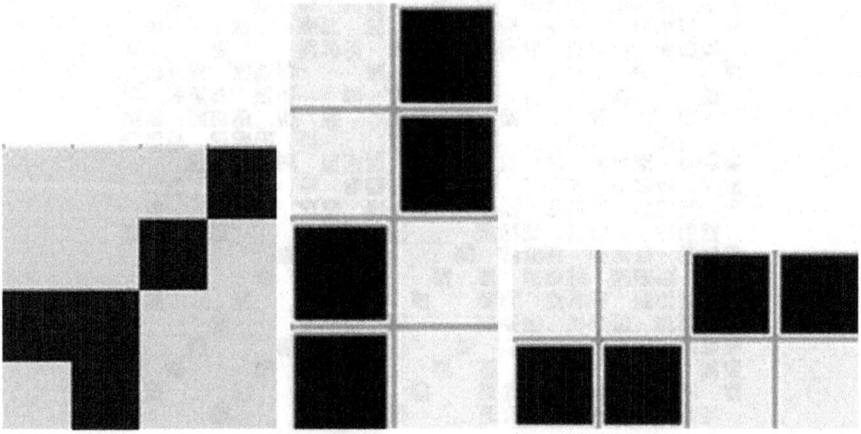

(a) (b) (c)

FIGURE 5.4 Figural rib: (a) motif; (b) warp rib $\frac{2}{2}$ as a basic pattern for the cells of the motif; and (c) weft rib $\frac{2}{2}$ as a basic pattern for the cells of the background.

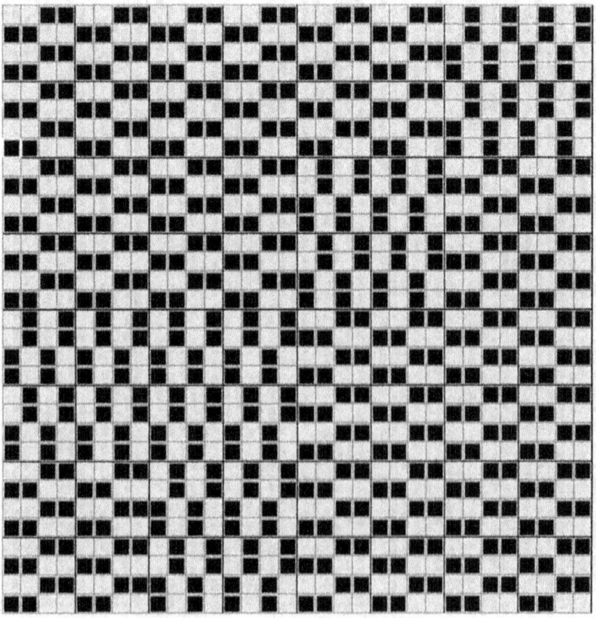

FIGURE 5.5 Figural rib based on a motif: the repeat.

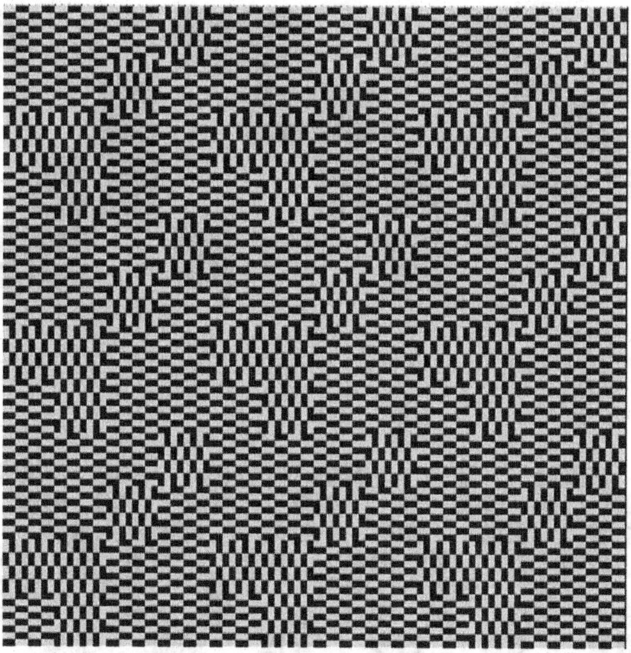

FIGURE 5.6 Cell-like weave based on a basket weave motif: the color design (black warp threads and white weft threads).

5.1.3 Fancy Twill

The fancy twill weave is designed based on a motif, whose cells are filled with right-handed twill and the background cells are filled with a left-handed twill. The basic twill weaves may be any elementary, strengthened, or complex twill weave with equal repeats and different diagonals (Z and S diagonals). The twill weave for the background cells starts with a thread, which overlaps are negative to the last thread of the twill weave for the motif.

Depending on the repeat of the basic weaves, the cells of the motif can be augmented one, two, or more times the repeat of the basic twills.

The warp repeat R_{wp} of the fancy twill is:

$$R_{wp} = A.R_{wp,m}.R_{wp,b} \tag{5.5}$$

where $A = 1,2,\ldots,n$ is the augmentation of each cell of the motif and $R_{wp,b}$ is the warp repeat of the basic twill weaves.

The weft repeat R_{wf} is:

$$R_{wf} = A.R_{wf,m}.R_{wf,b} \tag{5.6}$$

where $R_{wf,b}$ is the weft repeat of the basic twill weaves.

Figure 5.7 presents the motif for design of a fancy twill weave together with a twill weave $\frac{2}{.2}Z$ for the cells of the motif and the twill weave $\frac{2}{.2}S$ for the cells of

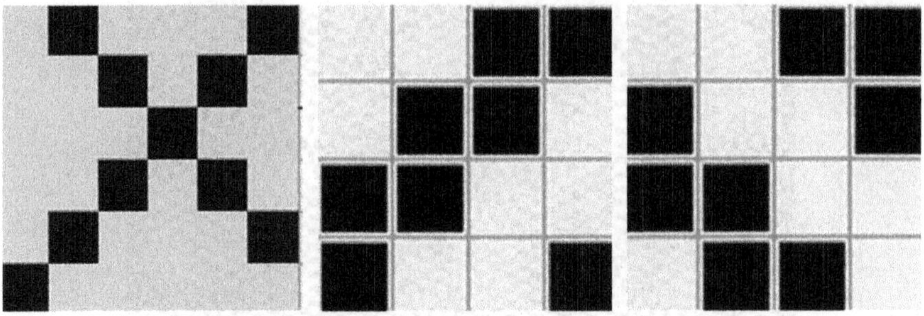

(a) (b) (c)

FIGURE 5.7 Fancy twill weave: (a) motif; (b) twill $\frac{2}{2}Z$ as a basic pattern for the cells of the motif; and (c) twill $\frac{2}{2}S$ as a basic pattern for the cells of the background.

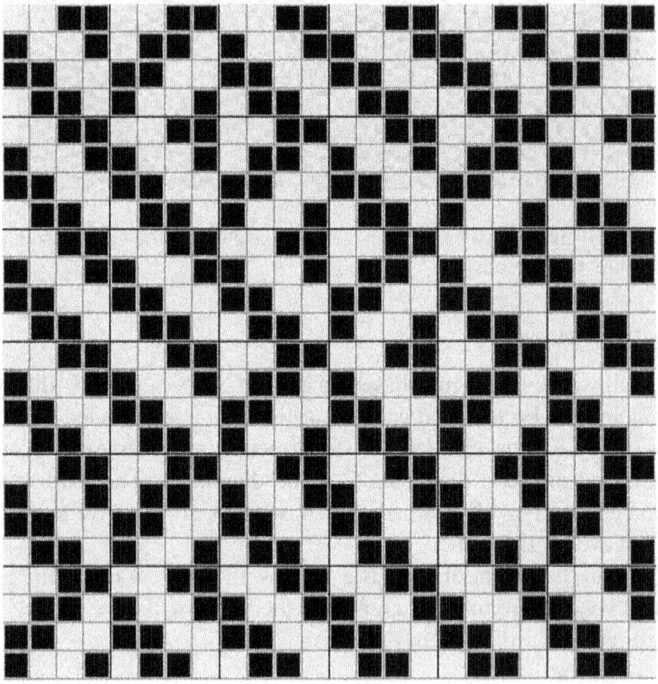

FIGURE 5.8 Fancy twill weave: the repeat.

the background. The selected augmentation is 1, which means that every cell of the motif will include a 4 × 4 net, equal to the repeat of the basic twills. The transition between the right-handed and left-handed twills is negative due to an appropriate start of the *S*-twill.

Figure 5.8 shows the repeat of the fancy twill weave. Figure 5.9 presents the color design of the fabric: black warp threads and white weft threads (three repeats are shown in both warp and weft directions).

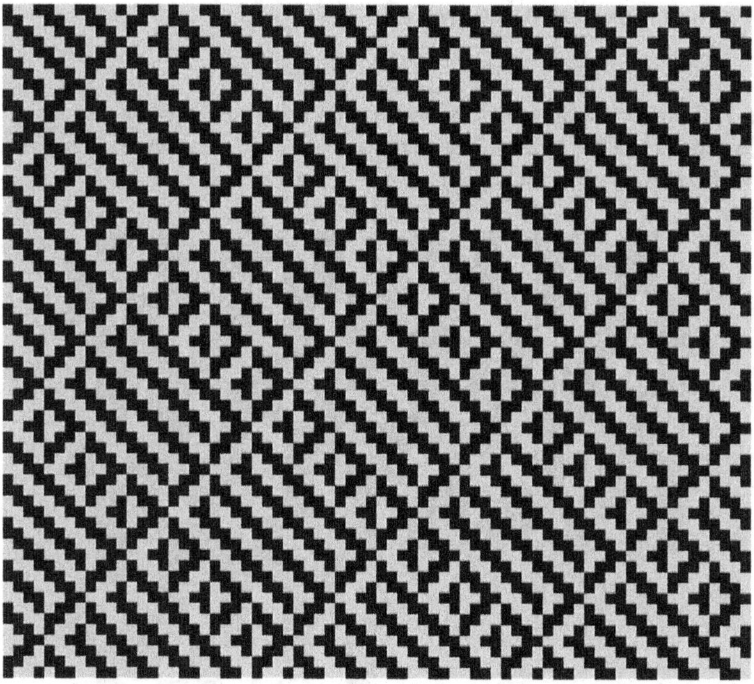

FIGURE 5.9 Fancy twill weave: the color design (black warp threads and white weft threads).

5.2 WEAVE PATTERNS FOR DIAGONAL EFFECTS

5.2.1 Diagonal Rib

The diagonal rib is designed when one repeat of a rib weave is set, following the order of a motif, which is an elementary weft-faced twill weave.

The basic weave is warp or weft rib and both its long and short repeats have to be multiples of the repeat of the motif. The displacement number for the whole rib weave is $S = +1$ if the motif is a right-handed twill and $S = -1$ if the motif is a left-handed twill weave.

If the basic weave is a warp rib, the warp repeat of the diagonal rib weave is calculated as:

$$R_{wp} = 2R_m, \tag{5.7}$$

where R_m is the square repeat of the motif.

The weft repeat of the diagonal rib weave is the same as the weft repeat of the warp rib.

If the basic weave is a weft rib, the warp repeat of the diagonal rib weave is equal to the warp repeat of the basic weave.

The weft repeat can be determined as:

$$R_{wf} = 2R_m \tag{5.8}$$

Figure 5.10 presents diagonal rib weave from a warp rib $\frac{3}{.2}$ as a basic weave and a motif, which is twill $\frac{1}{.4}Z$. Another diagonal rib weave is shown in Figure 5.11: the basic weave is weft rib $\frac{3}{.3}$ and the motif is twill weave $\frac{.1}{5.}S$.

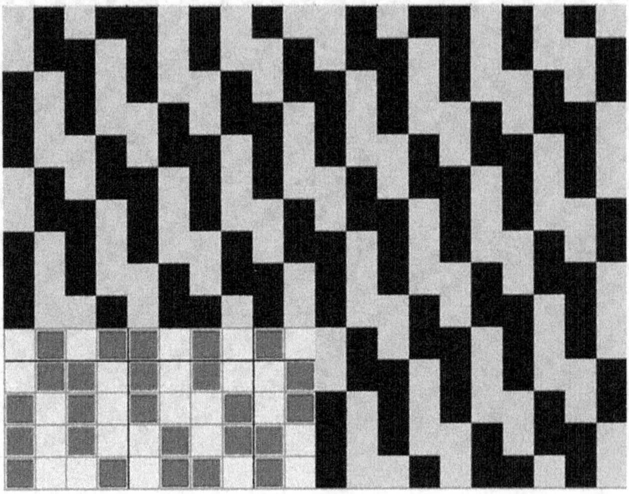

FIGURE 5.10 Diagonal rib weave from a warp rib $\frac{3}{.2}$ and a twill weave $\frac{1}{.4}Z$ as a motif.

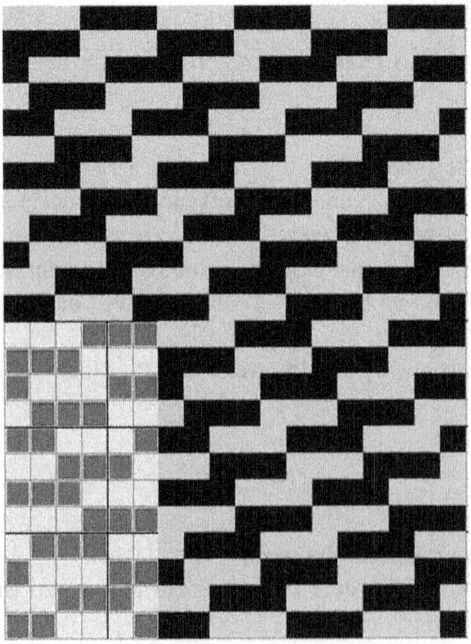

FIGURE 5.11 Diagonal rib weave from weft rib $\frac{3}{.3}$ and a twill weave $\frac{.1}{5.}S$ as a motif.

5.2.2 Broken Twill

The weave effect of the broken twill is two twill diagonals in the repeat (*Z* and *S*), both with a warp view or both with a weft view. An elementary or strengthened twill is usually used as a basic weave.

Broken twill weave is obtained by using a motif that usually is applied imaginarily: the repeat of the basic weave is divided into four quadrants. The twill diagonal occupies two of the opposite quadrants (Figure 5.12): I and III (*Z* twill) or II and IV (*S* twill). To design a broken twill weave, one of the twill diagonals rotates to an angle of 90°. Figure 5.13 shows a broken twill from twill $\frac{1}{.5}Z$.

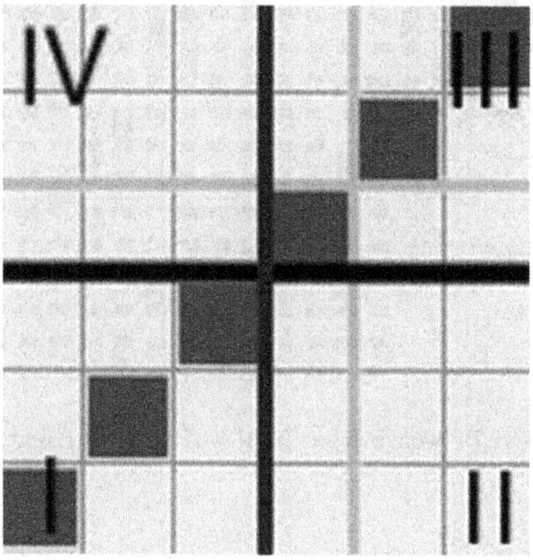

FIGURE 5.12 The imaginary motif of the broken twill.

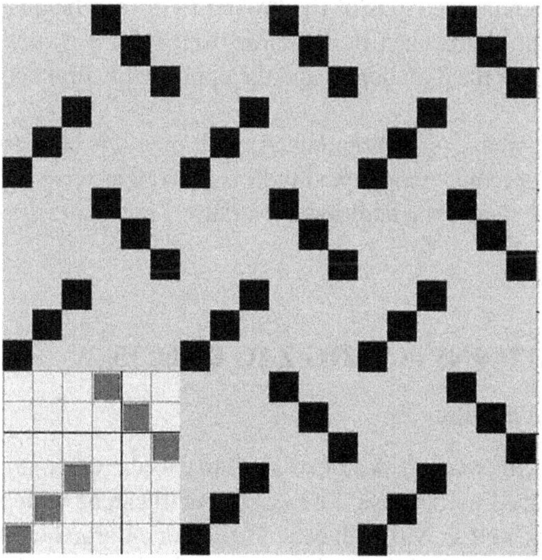

FIGURE 5.13 Broken twill weave from twill $\frac{1}{.5}Z$.

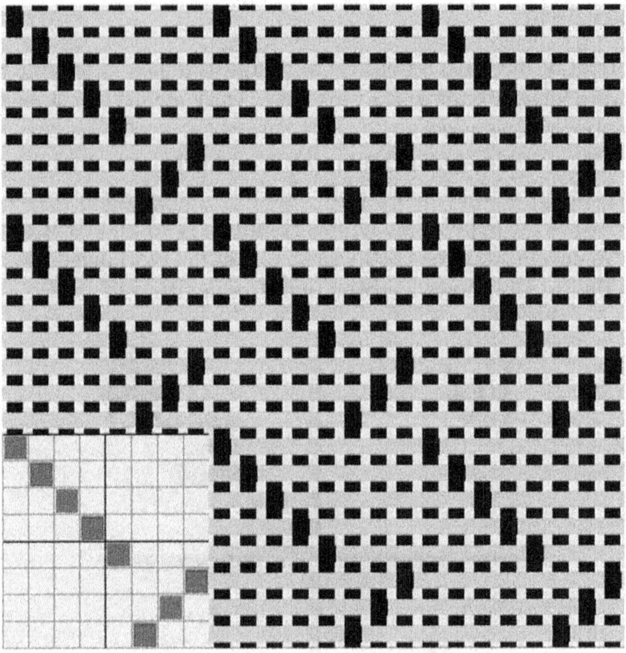

FIGURE 5.14 Broken twill weave from twill $\frac{1}{7}S$ with different size of the imaginary quadrants.

The size of the quadrants of the motif can be different. Thus, as long the repeat of the motif is, as many versions of the broken twill may be designed. Figure 5.14 illustrates an exemplary broken twill weave from twill $\frac{1}{7}S$: one of the quadrants involves 5 × 5 net (for the *S* diagonal) and the opposite quadrant consists of 3 × 3 net (the Z diagonal).

When the basic weave is a strengthened twill, then it is better to use the elementary twill from which the strengthened twill is derived as a basic weave for the broken twill, and later to add the additional overlaps. Figure 5.15 shows a broken twill from $\frac{2}{4}Z$.

5.3 WEAVE PATTERNS FOR ZIG-ZAG EFFECTS

5.3.1 Zig-Zag Warp Rib

The zig-zag warp rib weave is designed similarly to the diagonal rib, but the motif is a zig-zag weft-faced twill weave. The zig-zag twill can be with both one-thread or two-thread pick, as well as with reduced, normal, or enlarged teeth.

When the zig-zag effect of the motif is in the transversal direction of the cloth, a warp rib is chosen as the basic weave.

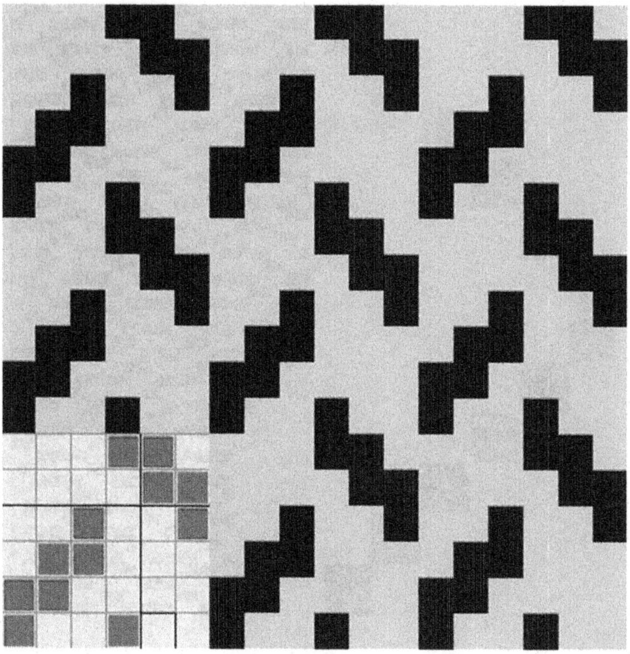

FIGURE 5.15 Broken twill weave from $\frac{2}{.4}Z$.

The warp repeat of the zig-zag warp rib can be calculated as:

$$R_{wp} = 2R_{wp,m} \tag{5.9}$$

The weft repeat of the zig-zag warp rib is equal to the weft repeat of the basic rib weave.

A zig-zag warp rib from a warp rib $\frac{3}{.3}$ and a motif (Figure 5.16a) is shown in Figure 5.16b. The motif is a zig-zag twill with one-thread pick from a weft-faced twill $\frac{1}{.4}Z$.

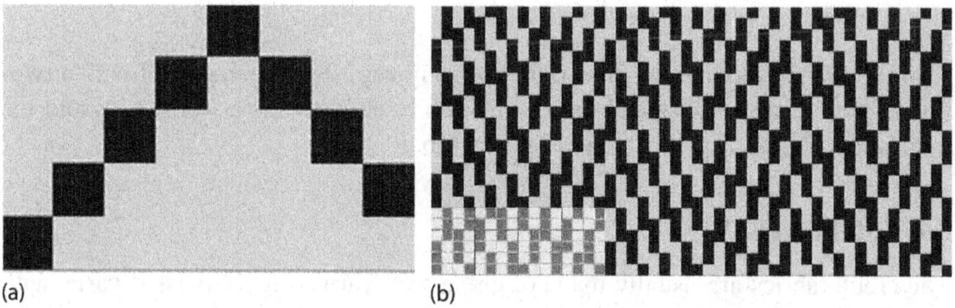

FIGURE 5.16 Transversal zig-zag rib weave: (a) motif – a zig-zag twill from a weft-faced twill $\frac{1}{.4}Z$ and (b) zig-zag rib weave.

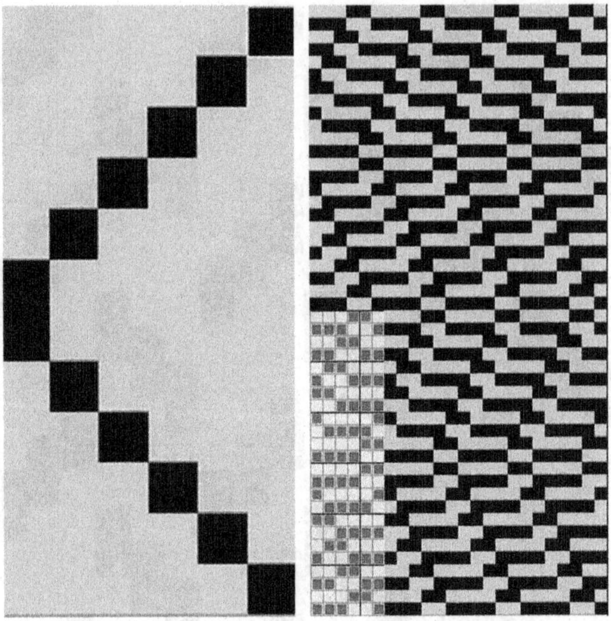

(a) (b)

FIGURE 5.17 Longitudinal zig-zag rib weave: (a) motif – a zig-zag twill from a weft-faced twill $\frac{\cdot 1}{5 \cdot}S$ and (b) zig-zag rib weave.

5.3.2 Zig-Zag Weft Rib

The zig-zag weft rib is designed when the effect of the motif is in the longitudinal direction of the cloth. Then, the basic weave for obtaining the pattern is a weft rib.

The warp repeat of the zig-zag weft rib is the same as the warp repeat of the basic weave. The weft repeat is determined as:

$$R_{wf} = 2R_{wf,m} \tag{5.10}$$

Figure 5.17a shows the motif for design of a zig-zag rib: a zig-zag twill with a two-thread pick from a weft-faced twill $\frac{\cdot 1}{5 \cdot}S$. The basic weave is a weft rib $\frac{4 \cdot}{\cdot 2}$ and the final weave design is presented in Figure 5.17b.

5.4 CREPE WEAVES

The crepe fabrics are usually made of crepe weft yarns: highly twisted yarns with different direction of the twist. There is another way to produce a crepe cloth: by using crepe weaves that create small figures or spots that are indiscernible for the eyes and create a sandy look of the fabric.

5.4.1 Crepe Weaves Based on Threads Rearrangement

This method for designing a crepe weave is based on the rearrangement of the threads of the basic weave. The scheme for threads rearrangement can be considered as a motif.

The basic weave is a complex twill weave, with at least four changes of the twill diagonal from warp to weft view and vice versa. The motif shows the new way for the arrangement of the threads: either on the warp or weft direction. The motif can be based on single threads, on groups of threads, or a combination of both.

Figure 5.18 shows the basic weave $\frac{4\,.\,2\,.}{.\,2\,.\,4}Z$, which is further used for designing of crepe weaves. The warp threads are 12, and the motif for their rearrangement is done following the order (in groups of two warp threads) 1-2-5-6-9-10-3-4-7-8-11-12. The result is shown in Figure 5.19.

The method does not always lead to the design of a crepe weave. Another motif of reordering the warp threads of the basic twill weave (again in groups of two warp threads): 11-12-9-10-7-8-5-6-3-4-1-2 results in a weave, in which the diagonal of the twill is perceivable (Figure 5.20). The designed weave is a version of a fancy twill, but it is not a crepe cloth.

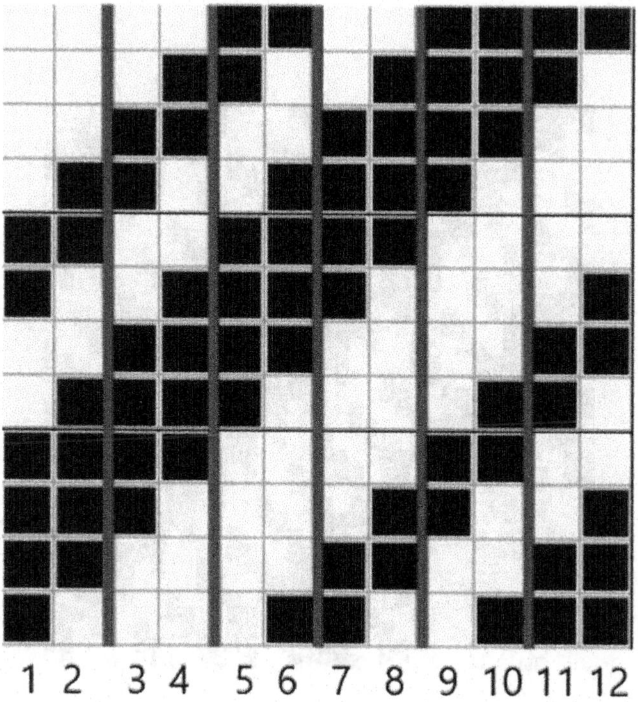

FIGURE 5.18 The basic weave $\frac{4\,.\,2\,.}{.\,2\,.\,4}Z$ for designing of a crepe weave.

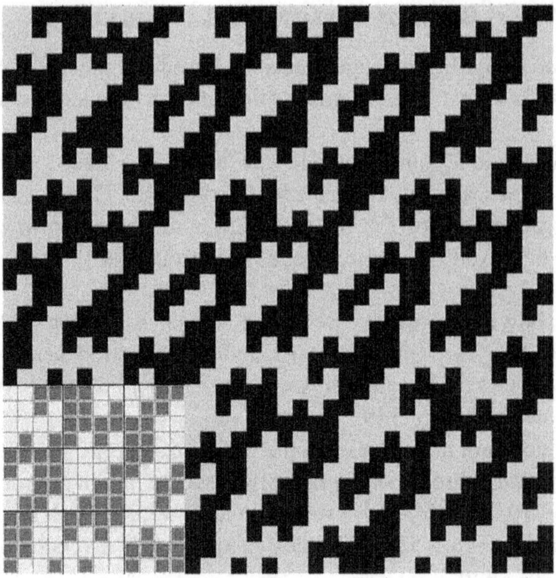

FIGURE 5.19 Crepe weave based on a motif for warp threads rearrangement 1-2-5-6-9-10-3-4-7-8-11-12 of the basic weave (Figure 5.18).

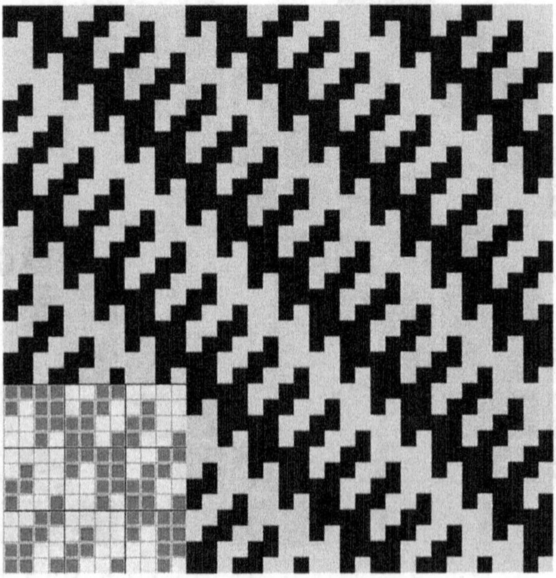

FIGURE 5.20 Fancy twill weave based on a motif for warp threads rearrangement 11-12-9-10-7-8-5-6-3-4-1-2 of the basic weave (Figure 5.18).

5.4.2 Crepe Weaves Based on a Motif and a Composition

A motif of small size and different shape is selected, together with a composition, which shows the position of the motif in the repeat of the designed weave. The motif can be a weave pattern, part of a weave pattern, or a free composition. The compositions itself can be of any type.

Figure 5.21a presents a composition, which requires the selected motif to rotate to 90°clockwise in each quadrant of the composition, from 1 to 4. The motif is shown in Figure 5.21b: a broken twill from twill $\frac{2}{2}Z$. The result from the joint effect of the motif and the composition is the crepe weave in Figure 5.22.

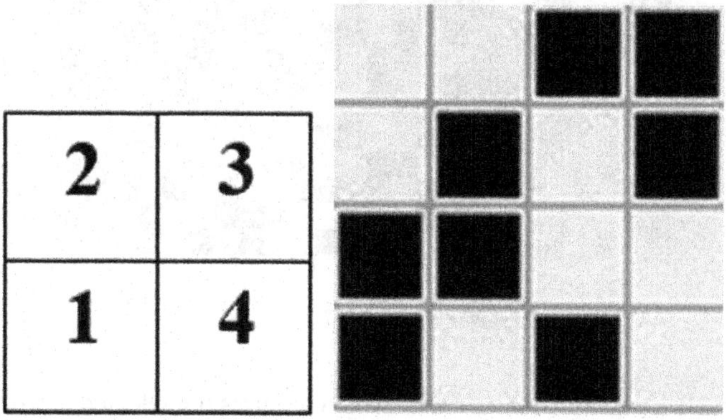

(a) (b)

FIGURE 5.21 Crepe weave based on a motif and a composition: (a) the composition for rotation of the motif to 90° clockwise from 1 to 4 and (b) the motif: a broken twill from twill $\frac{2}{2}Z$.

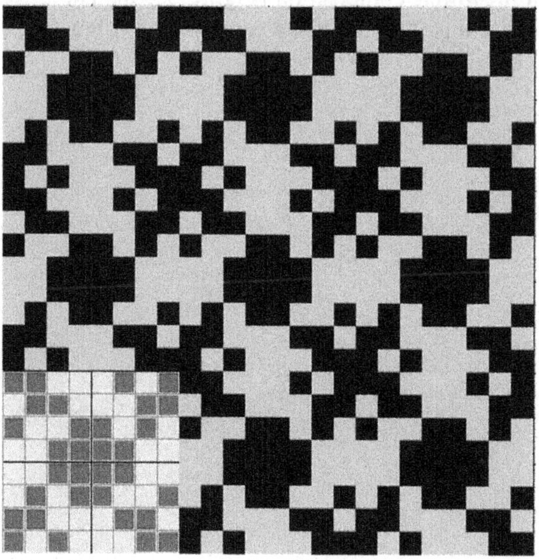

FIGURE 5.22 Crepe weave.

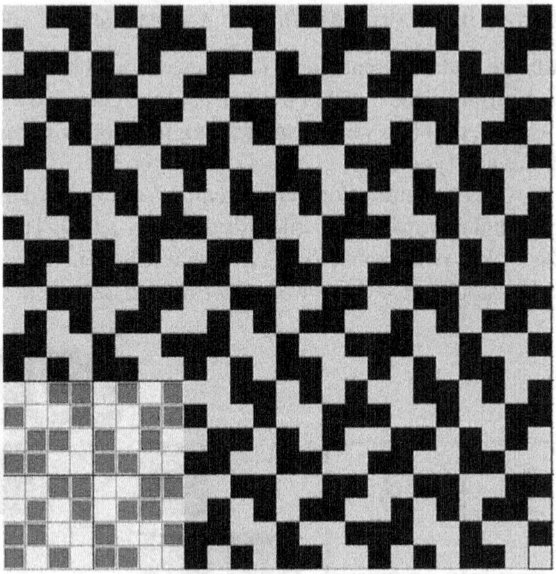

FIGURE 5.23 Version of the crepe weave.

Figure 5.21a presents a composition, which requires the selected motif to rotate to 90° clockwise in each quadrant of the composition, from 1 to 4. The motif is shown in Figure 5.21b: a broken twill from twill $\frac{2}{2}Z$. The result from the joint effect of the motif and the composition is the crepe weave in Figure 5.22.

Figure 5.23 shows a different crepe weave, designed from the same motif (Figure 5.21a) and a similar composition (Figure 5.21b). The difference is that the rotated motif in quadrants 2 and 4 of the composition (Figure 5.21b) are presented in their negative view.

6 New Weave Patterns Based on Patterns Insertion

6.1 WEAVE PATTERNS FOR RIB-LIKE EFFECTS

The warp and weft rib weaves result in horizontal or vertical rib texture of the cloth. The "ribs" are of limited length, however, as the long yarn floats lead to an unstable structure of the fabric.

The design of a new weave pattern with long floats requires additional overlaps to be foreseen between the warp and weft thread, thus assuring extra interlacing of the two sets of threads (warp and weft). The resulting tightened cloth structure is more stable, keeping the rib effect of the basic weave.

6.1.1 TIGHTENED RIB WITH ADDITIONAL OVERLAPS

The tightened rib with additional overlaps is known as a *soleil weave*. The basic weave is a rib weave with yarn float over six threads or more. The new interlacing points between the warp and weft threads are added, following the rules of an additional weave with small repeat and tight structure: plain weave, twill weave, hopsack weave, or rib weave.

The longer repeat of the basic rib weave and the repeat of the weave for the additional overlaps must be multiples to avoid needles augmentation of the final weave repeat.

The weave pattern, selected to strengthen the basic rib, is inserted according to the rule:

- as additional warp overlaps under the long warp floats of the warp rib;
- as additional weft overlaps under the long weft floats of the weft rib.

The warp repeat R_{wp} of the tightened warp rib with additional warp overlaps is calculated as:

$$R_{wp} = R_{wp,b} . R_{wp,s} \tag{6.1}$$

where $R_{wp,b}$ is the warp repeat of the basic weave (the warp rib) and $R_{wp,s}$ is the warp repeat of the weave, selected to strengthen the warp rib.

The weft repeat R_{wf} of the tightened warp rib with additional warp overlaps is:

$$R_{wf} = R_{wf,b} \tag{6.2}$$

where $R_{wf,b}$ is the weft repeat of the basic weave (the warp rib).

The warp repeat R_{wp} of the tightened weft rib with additional weft overlaps is calculated as:

$$R_{wp} = R_{wp,b} \tag{6.3}$$

where $R_{wp,b}$ is the warp repeat of the basic weave (the weft rib).

The weft repeat R_{wf} of the tightened warp rib with additional warp overlaps can be determined from:

$$R_{wf} = R_{wf,b}.R_{wf,s} \tag{6.4}$$

where $R_{wf,b}$ is the weft repeat of the basic weave (the weft rib) and $R_{wf,s}$ is the weft repeat of the weave, selected to strengthen the weft rib.

Figure 6.1 presents an exemplary tightened warp rib weave with additional warp overlaps. The basic weave is a warp rib $\frac{10.}{.10}$, strengthened with the plain weave. The weave effect on the cloth is horizontal ribs.

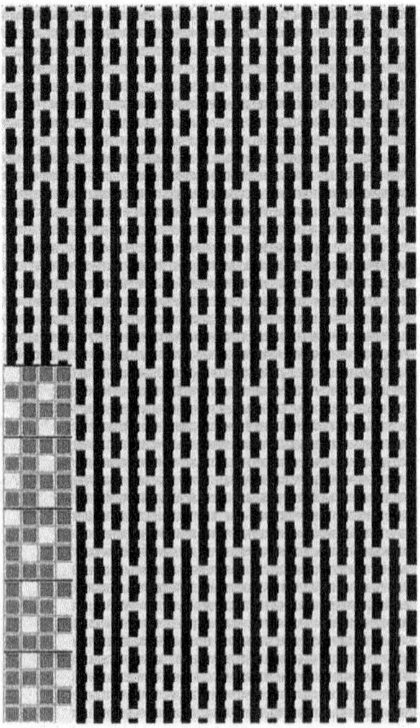

FIGURE 6.1 Tighten warp rib with additional warp overlaps from warp rib $\frac{10.}{.10}$ and plain weave.

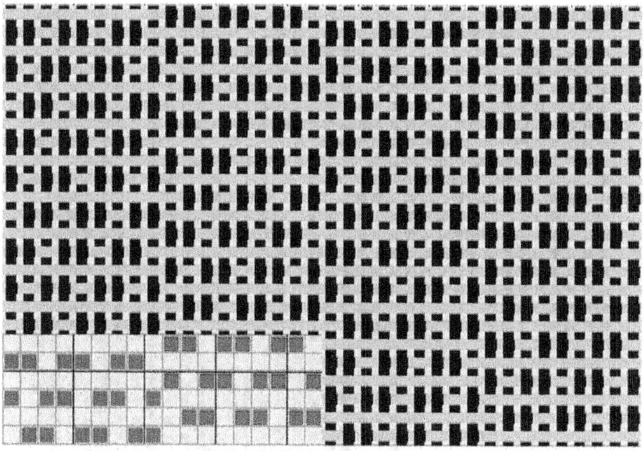

FIGURE 6.2 Tighten weft rib with additional weft overlaps from weft rib $\frac{9.}{.9}$ and twill weave $\frac{1.}{.2}Z$.

Figure 6.2 illustrates a tightened weft rib with additional weft overlaps. The basic weave is a weft rib $\frac{9.}{.9}$, strengthened following the rules of the twill weave $\frac{1.}{.2}Z$. The resulting weave effect is vertical ribs.

6.1.2 Tightened Rib with Additional Threads

The weave is very similar to the tightened rib with additional overlaps. The difference is that the threads for tightening the basic rib weave appear after one or more repeats of the rib weave. The ratio of alternation between the threads of the basic weave and the threads for tightening is 2:1 or 4:1.

The basic weave is a warp or weft rib with yarn float over six threads or more. The weave for tightening the rib weave has to be with a small repeat and compact structure: plain weave, twill weave, hopsack weave, rib weave, etc. The longer repeat of the rib and the repeat of the tightening weave have to be multiples.

Figure 6.3 illustrates the repeat of a tightened warp rib from a basic warp rib $\frac{10.}{.10}$ and a twill weave $\frac{1.}{.3}Z$, with warp yarn alternation of 2:1. Figure 6.4 shows the result of tightening the weft rib $\frac{7.}{.7}$ with the plain weave, following the weft yarn alternation 4:1.

6.1.3 Tightened Twill with Additional Weft Threads

The weaving effect of this weave is distinct embossed ribs in the direction of the weft threads. The horizontal ribs of the twill weave protrude, while the threads of the tightened weave form concaves.

The basic weave is an elementary twill with a repeat $R \geq 8$. The tightened weave is the plain weave. The alternation between the weft threads of the two weaves (basic to tightened) is 2:2 or 4:2.

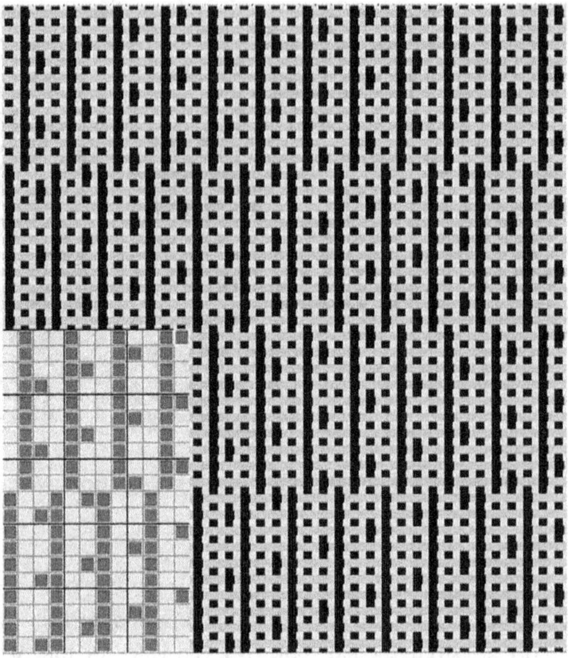

FIGURE 6.3 Tightened warp rib with additional threads from warp rib $\frac{10}{.10}$ and twill weave $\frac{1}{.3}Z$.

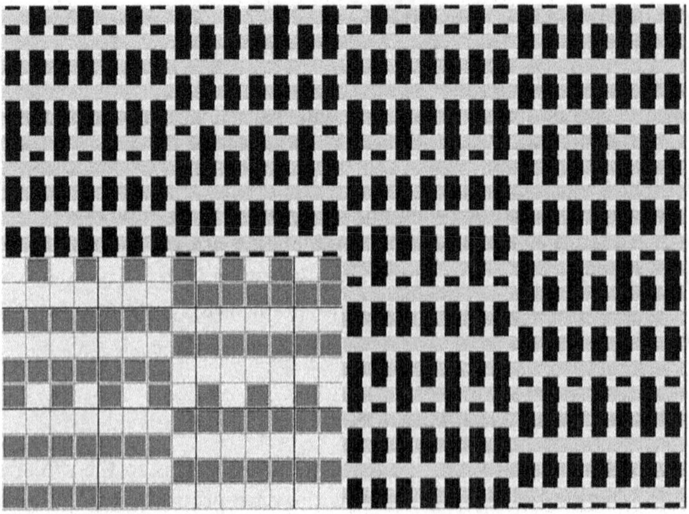

FIGURE 6.4 Tightened weft rib with additional weft threads from weft rib $\frac{7}{.7}$ and plain weave.

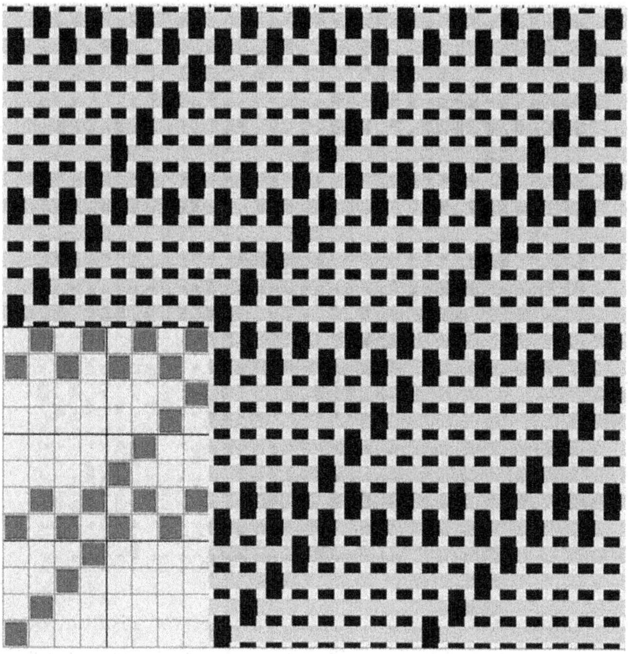

FIGURE 6.5 Tightened twill with additional weft threads from twill $\frac{1}{.7}Z$.

It is good the weft threads to be with a linear density two or more times higher than the warp threads to emboss the rib effect.

If the weft repeat of the new weave is too large, it can be shortened by using partly the basic twill weave, e.g., only the even or odd weft threads in the repeat.

An exemplary tightened twill with additional weft threads is shown in Figure 6.5. The basic weave is elementary twill $\frac{1.}{.7}Z$, tightened with the plain weave, applying an alternation 4:2.

6.1.4 Tightened Satin or Sateen with Additional Weft Threads

The texture of the tightened satin or sateen with additional weft threads involves embossed weft ribs, similar to these of the tightened twill with additional weft threads. The horizontal ribs of the satin or sateen weave protrude, while the ribs from the tightened weave drop.

The basic weaves are satin or sateen with a repeat $R \geq 8$. The weave for tightening is the plain weave. The alternation between the weft threads of the two weaves (basic to tightened) is 2:2 or 4:2. The use of tick weft threads (with two or more times higher linear density than the warp threads) leads to more visible rib effect.

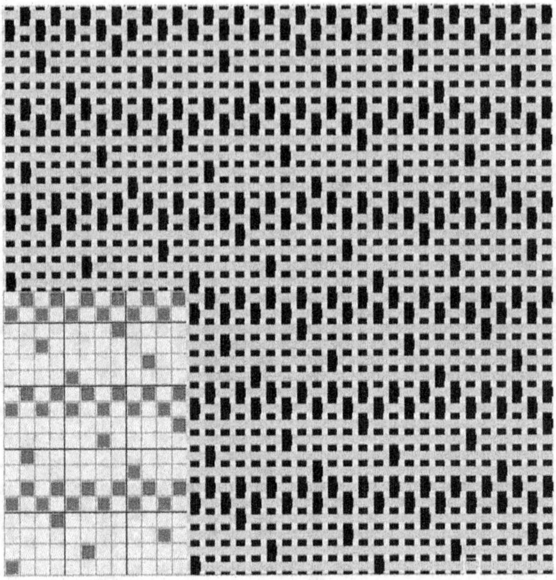

FIGURE 6.6 Tightened sateen weave with additional weft threads from sateen $\frac{1.}{.11}$ $(S = 5)$.

Figure 6.6 illustrates a tightened sateen weave with additional weft threads, designed from a sateen weave $\frac{1.}{.11}(S = 5)$ and the plain weave, applying an alternation between the weft threads 4:2.

If the weft repeat of the new weave is too large, it can be shortened, using the basic satin or sateen weave partly. Figure 6.7 shows the design of a tightened sateen by using the same basic weave from Figure 6.6, but only the odd weft threads are used, with an alternation with the weft threads of the plain weave 2:2.

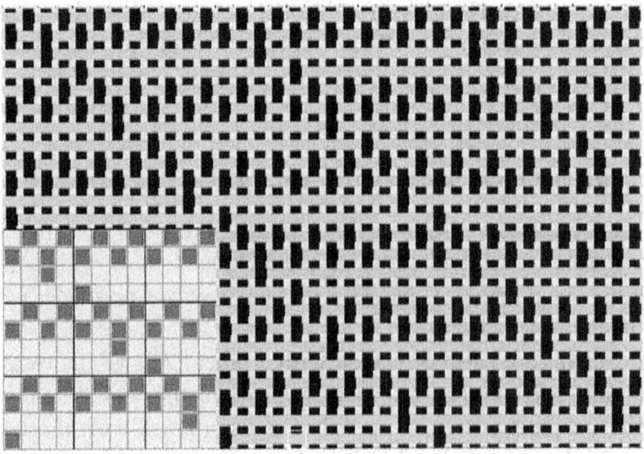

FIGURE 6.7 Tightened sateen weave with additional weft threads from the odd weft threads of sateen $\frac{1.}{.11}(S = 5)$.

6.2 WEAVE PATTERNS FOR DIAGONAL EFFECTS

6.2.1 Fancy Diagonal Twill by Warp Insertion

The weaving effect is obtained by inserting two or more twill weaves with the same direction of the diagonals: elementary, strengthened, complex, or fancy. The result is very similar to elongated twills obtained by changing the displacement number of the basic twill (see Chapter 7).

The repeats of the basic weaves should be equal or multiples. An alternation ratio between the threads of the basic weaves is used: most commonly 1:1.

When the insertion of the two basic weaves is done in the direction of the warp, the warp repeat can be, in most cases, calculated as:

$$R_{wp} = \sum_{i=1}^{k} n_{wp,i} \tag{6.5}$$

where $n_{wp,i}$ is the number of the warp threads in the repeat of the basic weave and k is the number of the basic twill weaves.

The weft repeat is calculated as the least common multiple (LCM) of the weft repeats of the basic twill weaves:

$$R_{wf} = \text{LCM}\left(R_{wf,b1}, R_{wf,b2}, \ldots, R_{wf,bk}\right) \tag{6.6}$$

where $R_{wf,bi}$ is the weft repeat of the single basic twill weave.

Figure 6.8 shows a fancy diagonal twill, designed by using the twill weaves $\frac{.3.2}{2.1.}S$ and $\frac{.3}{5.}S$, with an alternation of the warp threads 1:1.

6.2.2 Fancy Diagonal Twill by Weft Insertion

The requirements for the design of this weave are the same as for the fancy diagonal twill by warp insertion. The alternation ratio between the weft threads of the basic weaves is usually 1:1.

The warp repeat of the new weave is:

$$R_{wp} = \text{LCM}\left(R_{wp,b1}, R_{wp,b2}, \ldots, R_{wp,bk}\right) \tag{6.7}$$

where $R_{wp,bi}$ is the weft repeat of the single basic twill weave.

In most of the cases, the weft repeat can be calculated as:

$$R_{wf} = \sum_{i=1}^{k} n_{wf,i} \tag{6.8}$$

where $n_{wf,i}$ is the number of the weft threads in the repeat of the single basic twill weave.

Figure 6.9 shows a fancy diagonal twill by weft insertion, designed by using the same basic weaves as in Figure 6.8. The alternation of the weft threads is 1:1.

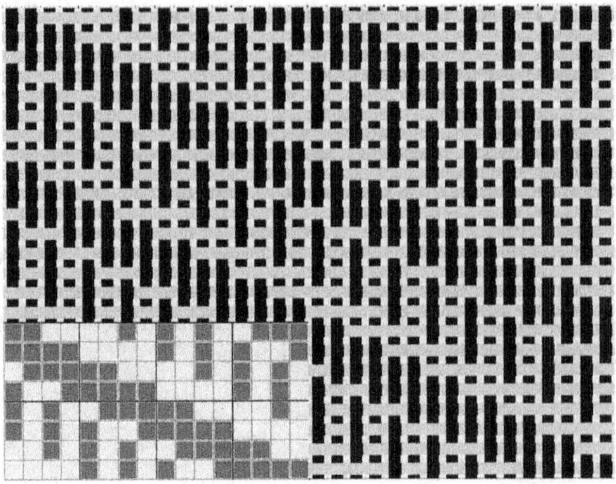

FIGURE 6.8 Fancy diagonal twill based on twills $\frac{.3.2}{2.1.}S$ and $\frac{.3}{5.}S$ (warp alternation 1:1).

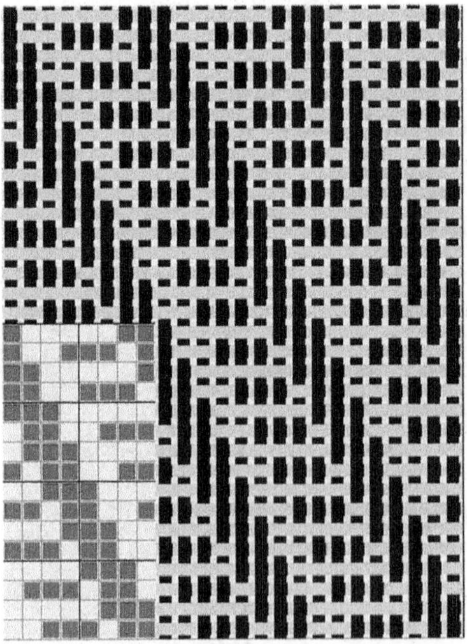

FIGURE 6.9 Fancy diagonal twill based on twills $\frac{.3.2}{2.1.}S$ and $\frac{.3}{5.}S$ (weft alternation 1:1).

6.3 CREPE WEAVES

The crepe weaves create a sandy look: the fabric seems to be covered by small grains or spots. The crepe weave imitates a cloth, woven by using crepe weft yarns that are highly twisted yarns with different direction of the twist.

6.3.1 Crepe Weave by Inserting Additional Overlaps

One of the methods to create a crepe weave is to randomly insert warp or weft overlaps in an existing weave pattern. The idea is to break the regularity of yarns' interlacing in the already known basic weave.

The basic weave can be a plain weave, rib, twill, satin, and other weaves with relatively small repeats. It is important to watch that no notable or symmetrical figures, diagonals, strips, large warp, or weft floats appear in the repeat of the crepe weave. Achieving these requirements is sometimes very difficult in a small repeat.

Figure 6.10a presents the repeat of a crepe weave, designed from a plain weave in a 6 × 6 net with two additional warp overlaps. The view of nine repeats of the crepe weave is illustrated in Figure 6.10b.

Figure 6.11a shows the repeat of a crepe weave from a weft rib weave $\frac{2\cdot}{\cdot 2}$ in an 8 × 8 net with eight additional warp overlaps. The designed crepe weave (nine repeats) is given in Figure 6.11b.

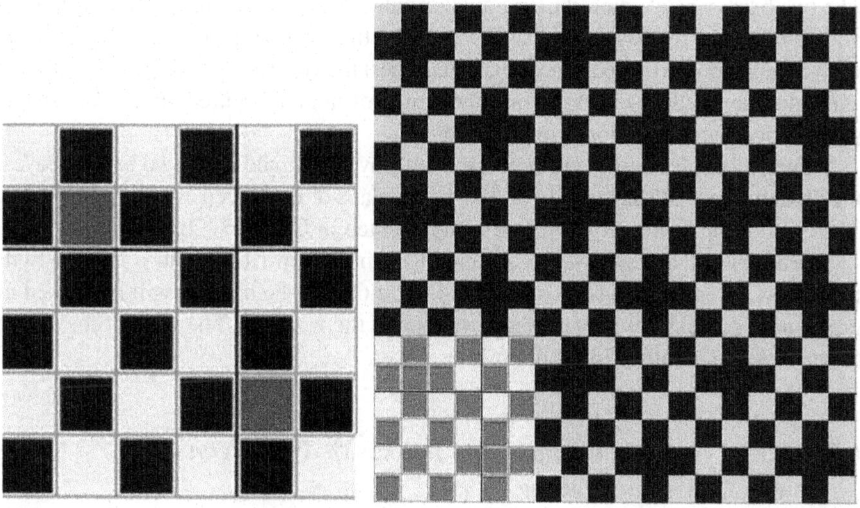

(a) (b)

FIGURE 6.10 Crepe weave from a plain weave in a 6 × 6 net: (a) repeat with marked additional warp overlaps and (b) colored design of black warp and white weft threads (3 × 3 repeats).

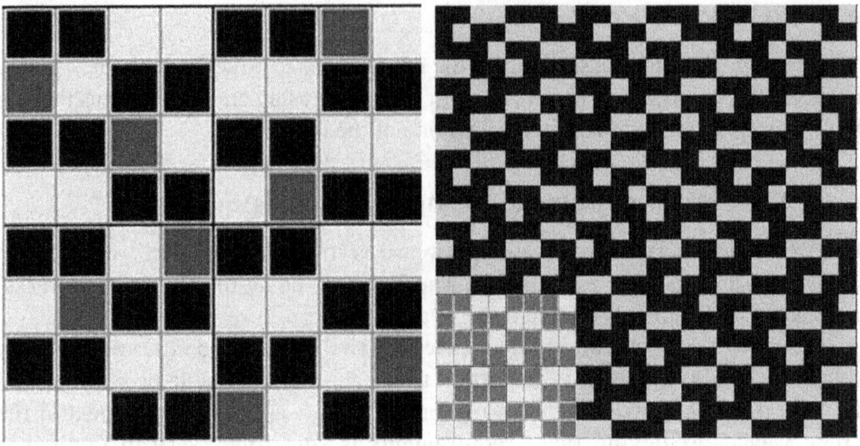

(a) (b)

FIGURE 6.11 Crepe weave from a weft rib $\frac{2\cdot}{\cdot 2}$ in an 8 × 8 net: (a) repeat with marked additional warp overlaps and (b) colored design of black warp and white weft threads (3 × 3 repeats).

6.3.2 Crepe Weave by Inserting One Weave over Another

The two basic weaves should have equal or multiple repeats. In the selected net (usually equal to the repeats of the basic weaves or the longer repeat, if one of the basic weaves has more extended repeat than the other) the two weaves are placed one over another. The result is checked for appearance of notable figures, diagonals, strips, symmetrical figures, or long threads float.

Figure 6.12a shows the repeat of a crepe weave, designed from two basic weaves: the plain weave and sateen $\frac{1\cdot}{\cdot 7}(S=3)$. The overlaps of the sateen are marked in gray. The view of the weave (3 × 3 repeats) is presented in Figure 6.12b.

The repeat of a crepe weave, designed from a warp rib $\frac{2\cdot}{\cdot 2}$ and a broken twill from a twill weave $\frac{1\cdot}{\cdot 3}Z$, is illustrated in Figure 6.13a. The broken twill is marked in gray, and it coincides with two warp overlaps of the warp rib. The view of the weave (3 × 3 repeats) is shown in Figure 6.13b.

6.3.3 Crepe Weave by Inserting the Threads of Two Weaves in a Given Relationship

The crepe weave is designed by using two (usually not more) basic weaves with equal or multiple repeats. The ratio of threads insertion is 1:1, 2:1, 1:2, or 2:2 threads. Both warp and weft threads can be subjects of these alternations.

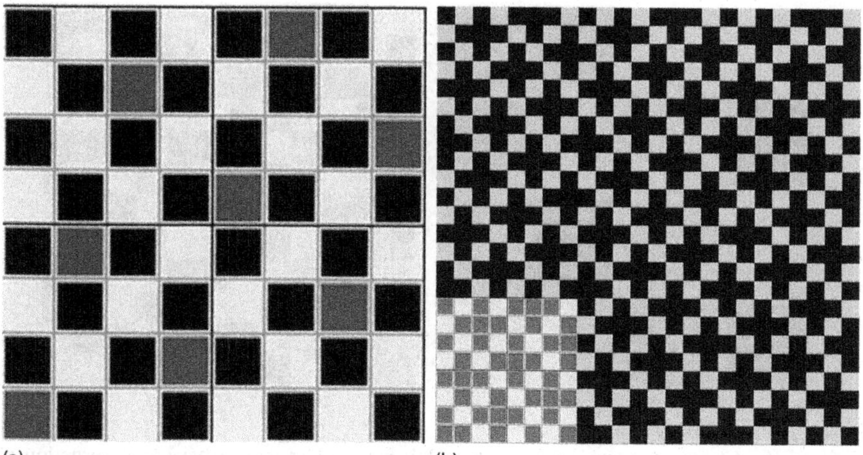

(a) (b)

FIGURE 6.12 Crepe weave from the plain weave and sateen $\frac{1\cdot}{\cdot 7}(S=3)$: (a) repeat with marked overlaps of the sateen weave and (b) colored design of black warp and white weft threads (3 × 3 repeats).

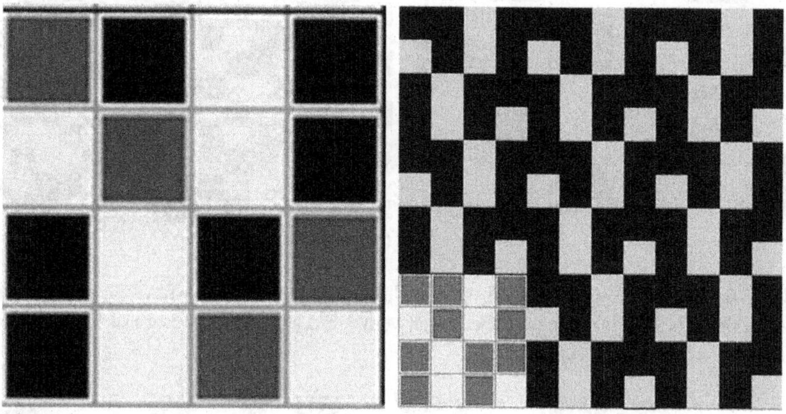

(a) (b)

FIGURE 6.13 Crepe weave from a warp rib $\frac{2\cdot}{\cdot 2}$ and a broken twill from a twill weave $\frac{1\cdot}{\cdot 3}Z$: (a) repeat with marked overlaps of the broken twill and (b) colored design of black warp and white weft threads (3 × 3 repeats).

Figures 6.14 through 6.16 show three crepe weaves, all designed with the plain weave and warp weave $\frac{2\cdot}{\cdot 2}$ as basic weaves. The difference between the new crepe weaves is the alternation of the warp threads: 1:1 (Figure 6.14a), 2:1 (Figure 6.14b), and 2:2 (Figure 6.14c). All warp rib overlaps are marked in gray. The designed fabrics for the different warp threads alternations are presented in Figures 6.15b, 6.16b.

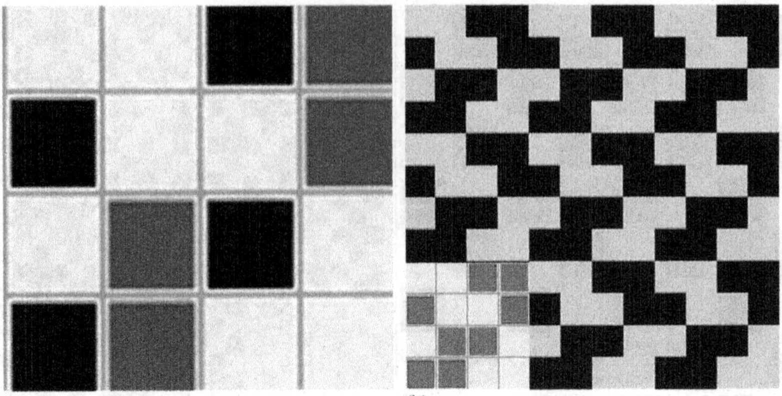

(a) (b)

FIGURE 6.14 Crape weave based on the plain weave and warp weave $\frac{2\cdot}{\cdot 2}$ (alternation of the warp threads 1:1): (a) repeat of the crepe weave and (b) colored design of black warp and white weft threads (3 × 3 repeats).

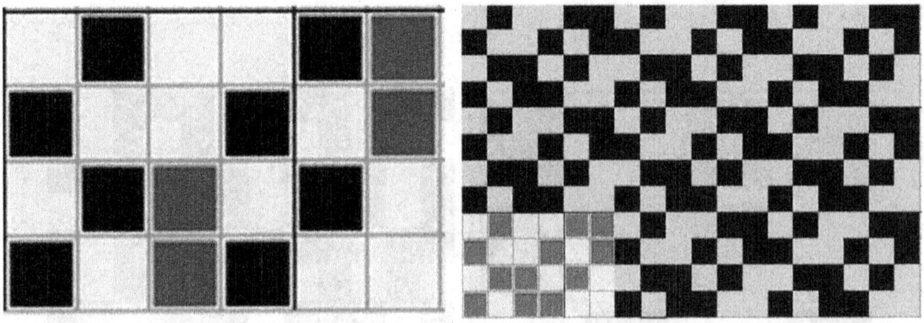

(a) (b)

FIGURE 6.15 Crape weave based on the plain weave and warp weave $\frac{2\cdot}{\cdot 2}$ (alternation of the warp threads 2:1): (a) repeat of the crepe weave and (b) colored design of black warp and white weft threads (3 × 3 repeats).

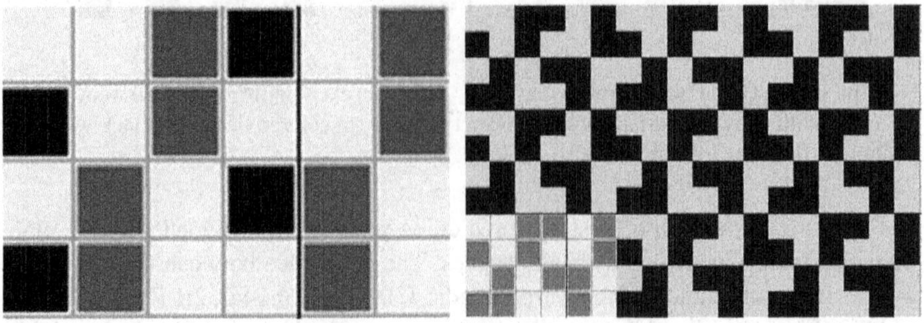

(a) (b)

FIGURE 6.16 Crape weave based on the plain weave and warp weave $\frac{2\cdot}{\cdot 2}$ (alternation of the warp threads 1:2): (a) repeat of the crepe weave and (b) colored design of black warp and white weft threads (3 × 3 repeats).

7 New Weave Patterns Based on the Change of the Displacement Number

7.1 WEAVE PATTERNS FOR DIAGONAL EFFECTS

In principle, the twill weaves (elementary twill, strengthened twill, and complex twill) create a diagonal effect on the cloth surface. When the warp and weft densities of the fabric are equal, the diagonal runs at an angle of 45°.

One way to change the degree of the twill diagonal is to change the ratio between the warp and weft densities. Another way (when the densities in both directions are the same) is to use special twill weaves, *elongated twills*, with changed displacement number, i.e., $S \neq \pm 1$.

Two types of elongated twills can be designed this way (Figures 7.1 and 7.2):

- Steep twills: when the angle of the twill diagonal is bigger than 45°.
- Reclined twills: when the angle of the twill diagonal is lower than 45°.

To design an elongated twill weave, a basic twill weave must be selected together with a displacement number S, different from 1 (as in the case of satin and sateen weaves). The displacement number should not break the largest twill diagonal; therefore, S can run between 2 and the largest twill diagonal minus one overlap (i.e., for the basic twill weave $\frac{6\cdot}{\cdot 3}Z$, the displacement number $S = 2 \div 5$). Besides, it is reasonable S to be a multiple of the repeat of the basic weave; otherwise, the repeat of the designed elongated twill weave could be quite large.

The displacement number can be indicated in the formula of the twill repeat:

- As an exponent: in the case of the steep twills, the displacement number counts upwards.
- As an index: in the case of the reclined twills, usually, the indicated overlaps are placed over the first weft thread, and the displacement number counts to the right.

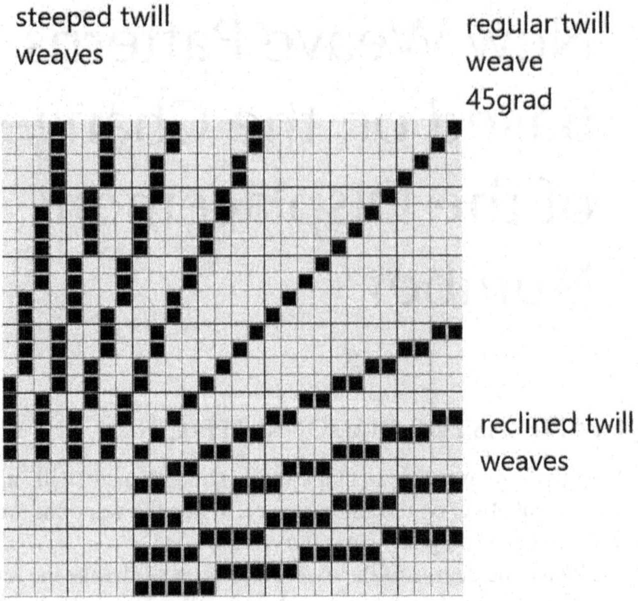

FIGURE 7.1 Elongated twills.

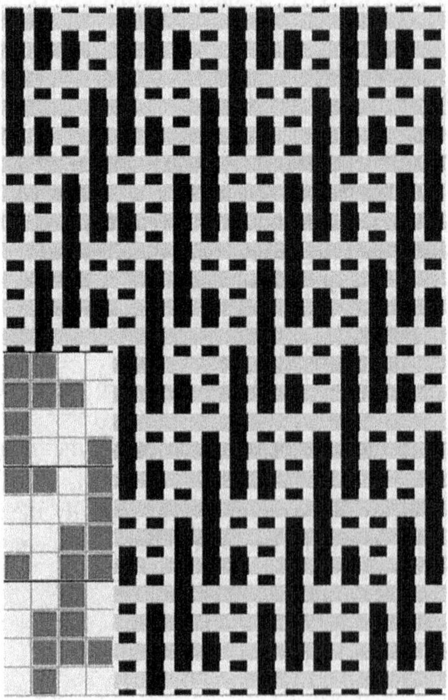

FIGURE 7.2 Steep twill weave from a complex twill weave $\frac{1.\ 5}{4.\ 2.}Z^3$.

7.1.1 Steep Twill

The warp repeat of the steep twill R_{wp} is calculated from:

$$R_{wp} = \frac{R_{wp,b}}{S} \tag{7.1}$$

where $R_{wp,b}$ is the warp repeat of the basic twill weave.

The weft repeat R_{wf} of the steep twill is the same as the weft repeat of the basic twill $R_{wf,b}$.

7.1.2 Reclined Twill

The warp repeat R_{wp} of the reclined twill is equal to the warp repeat of the basic twill $R_{wp,b}$.

The weft repeat of the reclined twill R_{wf} will be:

$$R_{wf} = \frac{R_{wf,b}}{S} \tag{7.2}$$

where $R_{wf,b}$ is the weft repeat of the basic twill weave.

A reclined twill weave from a strengthened twill weave $\frac{6.}{.4}Z_2$ is shown in Figure 7.3.

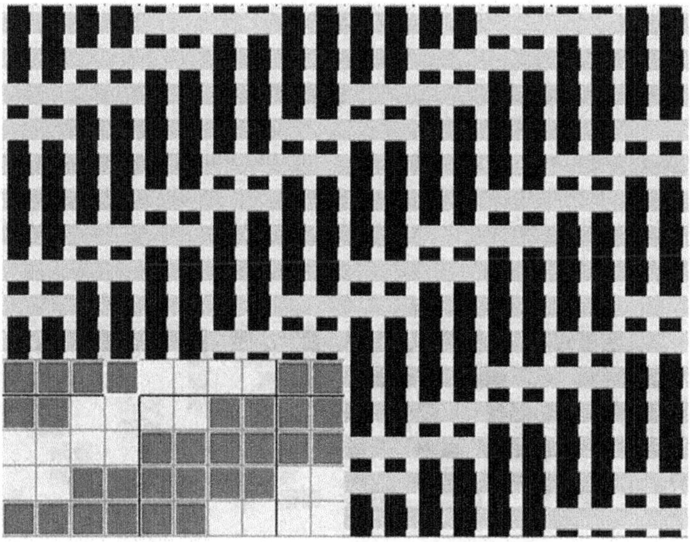

FIGURE 7.3 Reclined twill weave from a strengthened twill weave $\frac{6.}{.4}Z_2$.

7.2 WEAVE PATTERNS FOR CURVED EFFECTS

The curved twill weave creates a curvilinear diagonal effect transversely or longitudinally on the cloth. The pattern of the weave is designed within several consecutive repeats of the basic weave by changing the displacement number in each repeat.

The basic weave for creation of the curved twill is strengthened twill weave, with a warp-faced diagonal of at least three warp overlaps. The curved twill weave is designed within several stages, which may involve a different number of threads, but usually, each stage is equal to the repeat of the basic weave.

The displacement number changes from $S = 1$ in the first stage, $S = 2$ in the second stage until reaching S_{max}; subsequently, the displacement number decreases from S_{max} to $S = 2$.

The maximal displacement number S can be calculated from the number of the warp overlaps n_o in the diagonal of the basic twill weave:

$$S_{max} = n_o - 1 \tag{7.3}$$

When the maximal displacement number S_{max} is used, the curvilinear weave effect is also maximal. If this is not the goal of the design, a lower maximal displacement number S_r may be used, i.e., $S_r < S_{max}$. The obtained curvilinear effect will be lighter.

The number of the stages P depends on the desired curvilinear effect on the cloth: a full curve or lighter version. In the case of a full curvilinear effect, the number of the necessary stages is calculated as:

$$P = 2S_{max} - 2 \tag{7.4}$$

If the goal is a lighter curve on the cloth, the number of stages is:

$$P = 2S_r - 2 \tag{7.5}$$

7.2.1 LONGITUDINALLY CURVED TWILL WEAVE

The design of a curved twill weave for longitudinal effects from a basic twill weave $\frac{4}{.4}Z$ is shown in Figures 7.4 through 7.7. A full curvilinear effect is obtained, applying the maximal displacement number $S_{max} = 4 - 1 = 3$. The number of stages is 4.

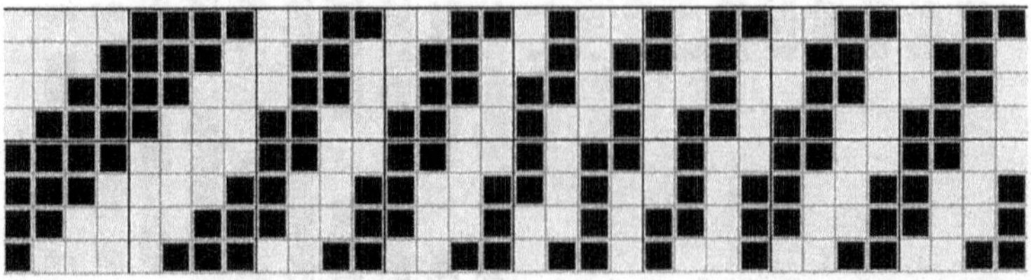

FIGURE 7.4 Longitudinally curved twill weave from a twill $\frac{4}{.4}Z$: the repeat (eight threads for a stage).

Figure 7.4 presents the repeat of the longitudinally curved twill if a whole repeat of the basic weave is used (eight threads). A full impression of the designed weave can be obtained from Figure 7.5, where two warp and ten weft repeats of the curved twill are shown.

The decrement of the number of the warp threads in a stage can lead to a different view of the final weave. Figure 7.6 shows the repeat of the longitudinally

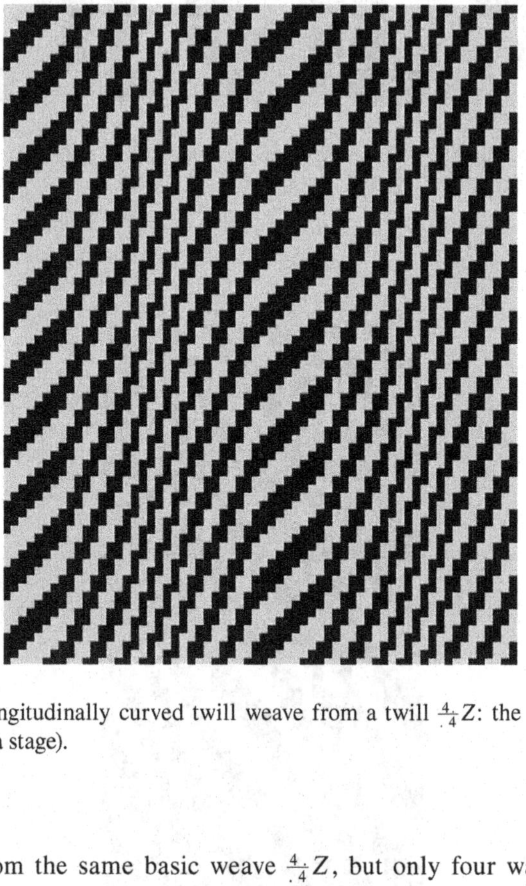

FIGURE 7.5 Longitudinally curved twill weave from a twill $\frac{4}{.4}Z$: the view of the cloth (eight threads for a stage).

curved twill from the same basic weave $\frac{4}{.4}Z$, but only four warp threads are used in each stage. The warp repeat of the final weave, as shown in Figure 7.7, is shorter and the curve line is different, compared to Figure 7.6.

7.2.2 Transversally Curved Twill Weave

The repeat of a transversally curved twill weave from the same basic twill weave $\frac{4}{.4}Z$ is given in Figure 7.8. A full curvilinear effect is obtained, applying the maximal displacement number $S_{max} = 3$ together with a whole repeat of the basic weave (eight threads) for each of the stages (four stages). The impression of the designed curved twill is given in Figure 7.9, where ten warp and two weft repeats of the weave are presented.

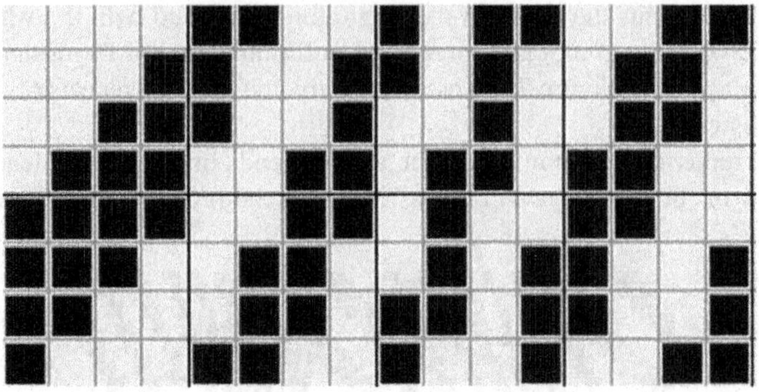

FIGURE 7.6 Longitudinally curved twill weave from a twill $\frac{4}{\cdot 4}Z$: the repeat (four threads for a stage).

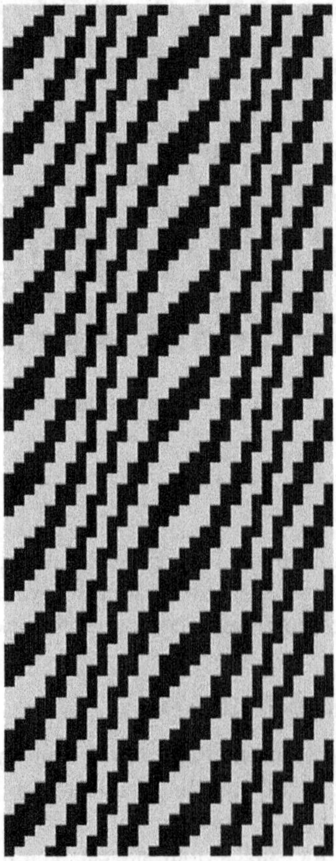

FIGURE 7.7 Longitudinally curved twill weave from a twill $\frac{4}{\cdot 4}Z$: the view of the cloth (four threads for a stage).

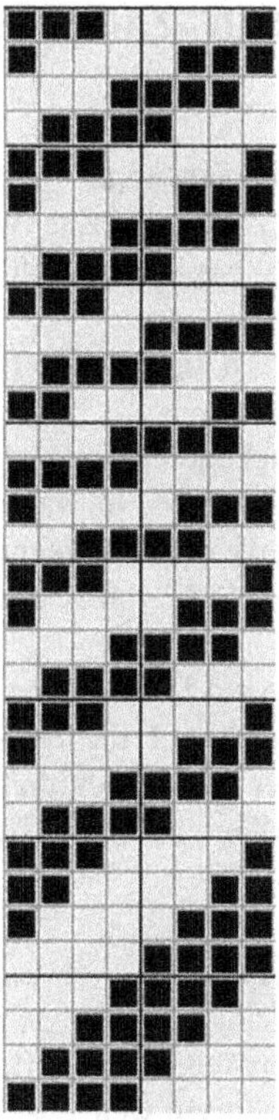

FIGURE 7.8 Transversally curved twill weave from a twill $\frac{4}{4}Z$: the repeat (eight threads for a stage).

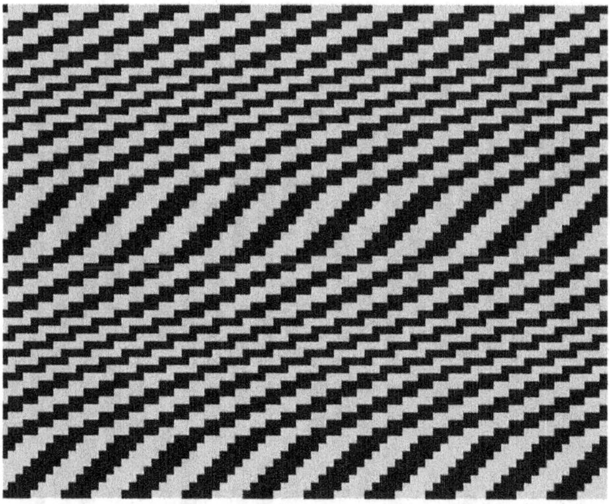

FIGURE 7.9 Transversally curved twill weave from a twill $\frac{4}{4}Z$: the view of the cloth (eight threads for a stage).

7.3 WEAVE PATTERNS FOR ZIG-ZAG EFFECTS

7.3.1 Herringbone Twill

The herringbone twill is a further version of the zig-zag twill: the displacement number *S* of the basic weave changes to form an opposite twill diagonal – from *Z* to *S* or vice versa. The face of the diagonal changes together with the displacement number – from warp face to weft face or vice versa.

The weave effect on the cloth is a zig-zag line from diagonals with shifting warp and weft effects.

The basic weave for the herringbone twill may be strengthened, complex or elongated twill weaves with an equal number of warp n_{wp} and weft n_{wf} overlaps in the repeat. After several threads, both the displacement number *S* and the view of the diagonal undergo changes. The number of the threads may be lower, equal, or higher than the weave repeat *R*.

Figures 7.10 through 7.12 show a herringbone twill weave from a basic weave $\frac{3}{3}Z$. In the first weave, the change of the diagonal appears after four threads (Figure 7.10); in the second, after six threads that is the same as the repeat of the basic weave (Figure 7.11). The diagonal of the basic twill in Figure 7.12 changes after eight threads.

7.3.2 Curved Zig-Zag Twill Weave

The curved zig-zag twill weave creates a wavy effect on the cloth. The same design method is applied as for the curved twill weave. After the design of the repeat of the curved twill, the mirror image of the repeat is added to obtain the repeat of the curved zig-zag twill.

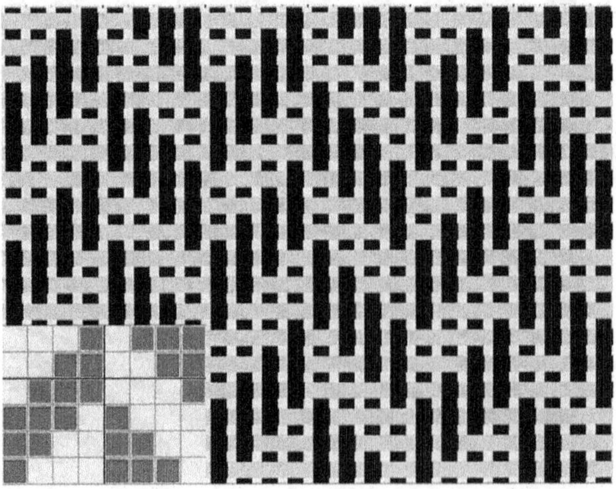

FIGURE 7.10 Herringbone twill weave from a strengthened twill $\frac{3}{3}Z$ and diagonal change after four threads.

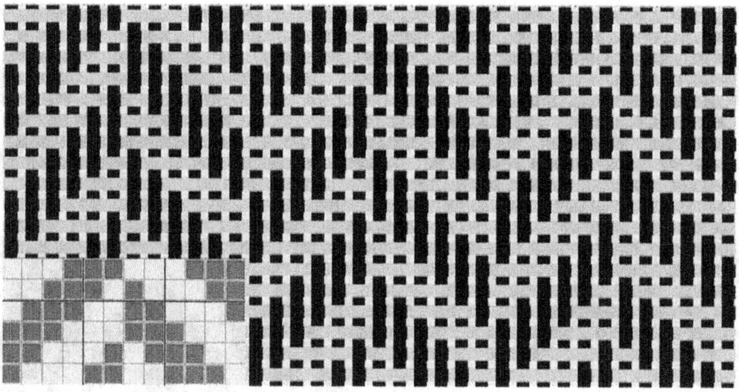

FIGURE 7.11 Herringbone twill weave from a strengthened twill $\frac{3}{3}Z$ and diagonal change after six threads.

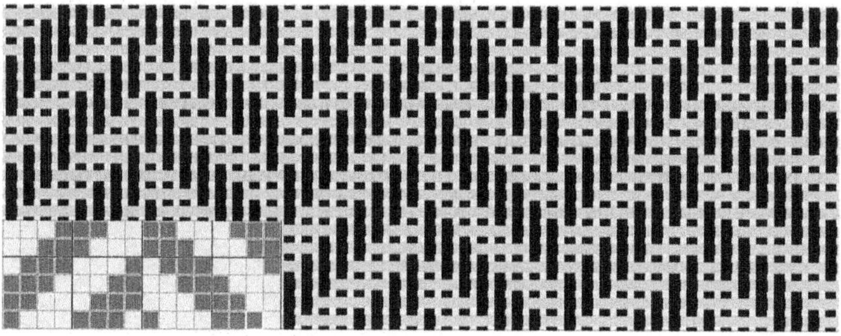

FIGURE 7.12 Herringbone twill weave from a strengthened twill $\frac{3}{3}Z$ and diagonal change after eight threads.

Thus, the curved zig-zag twill is designed again by using changes in the displacement number S in each of the stages. The stage can involve a different number of threads, depending on the goal: more threads in the stage increase both the thickness of the curve line and the repeat of the final weave.

The number of stages P in the repeat can be calculated as:

$$P = 2(2S_{max} - 1) \tag{7.6}$$

where S_{max} is the maximal displacement number, or:

$$P = 2(2S_r - 1) \tag{7.7}$$

where S_r is the applied maximal displacement number for obtaining partial curvature, which is lower than S_{max}.

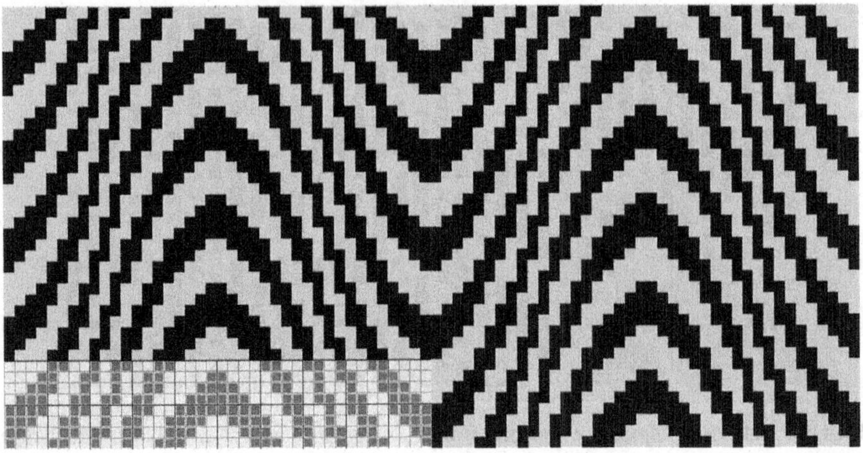

FIGURE 7.13 Curved zig-zag twill weave from a twill $\frac{4}{.4}Z$ (four threads for a stage).

Figure 7.13 shows a curved zig-zag twill weave from a twill $\frac{4}{.4}Z$, using four warp threads for each stage. Adding the mirror image of the curved twill (Figure 7.7) results in picks of two threads in the final repeat of the curved zig-zag twill (Figure 7.13). If this is not the intention, the repeat of the curved zig-zag twill can be reduced by one thread to form a single-thread pick.

Section II

New Methods for Design of New Weave Patterns

8 New Weave Patterns Based on Boolean Operations

8.1 THE CONCEPT OF THE METHOD

The weave diagrams are symbolic representations of the interlacement between the two sets of perpendicular threads: warp set and weft set. From the point of view of the mathematics of logic, the weave pattern is a combination of Boolean variables, which can take only two values: true or false. As a result, the warp and weft overlaps in the weave repeat have the following logical meaning:

- Warp overlap: "it is true that the warp thread is on the top of the cloth" or "it is false that the weft thread is on the top of the cloth".
- Weft overlap: "it is true that the weft thread is on the top of the cloth" or "it is false that the warp thread is on the top of the cloth".

The method was proposed first in the research by Griswold (2002). By the analogy with the Boolean algebra, the warp and weft overlaps can be treated as logical variables with two possible values: 0 or 1. Practically, every weave pattern uses a binary code: Figure 8.1 shows the repeat of a broken twill weave from a basic twill $\frac{4}{.4}Z$ (Figure 8.1), presented in binary code (Figure 8.2b).

The analysis of the 16 Boolean operations, made by Griswold (2002), shows that the application of only part of them leads to the design of new weave patterns. These operations over the two variables X and Y are:

- Negation (or not) : $\sim X$ or $\sim Y$
- Disjunction : $X + Y$
- Conjunction : $X \times Y$
- Exclusive or : $X \oplus Y$
- Equivalence : $X \leftrightarrow Y$
- Implication : $X \rightarrow Y$

It was proven that an additional operation Implication: $Y \rightarrow X$ can be successfully applied for designing a new weave (Angelova, 2005). This operation is analogous to Implication: $X \rightarrow Y$ and has a similar meaning: if Y implies X and Y is true, then X is true.

The logical relationship between the values of the variables X and Y, according to the Boolean operations, is shown in Table 8.1.

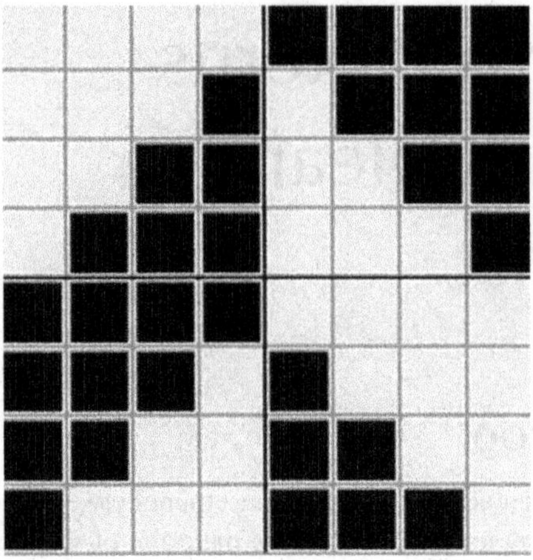

(a)

0	0	0	0	1	1	1	1
0	0	0	1	0	1	1	1
0	0	1	1	0	0	1	1
0	1	1	1	0	0	0	1
1	1	1	1	0	0	0	0
1	1	1	0	1	0	0	0
1	1	0	0	1	1	0	0
1	0	0	0	1	1	1	0

(b)

FIGURE 8.1 Broken twill weave from a basic twill $\frac{4}{4}Z$: (a) the repeat and (b) the repeat in a binary code.

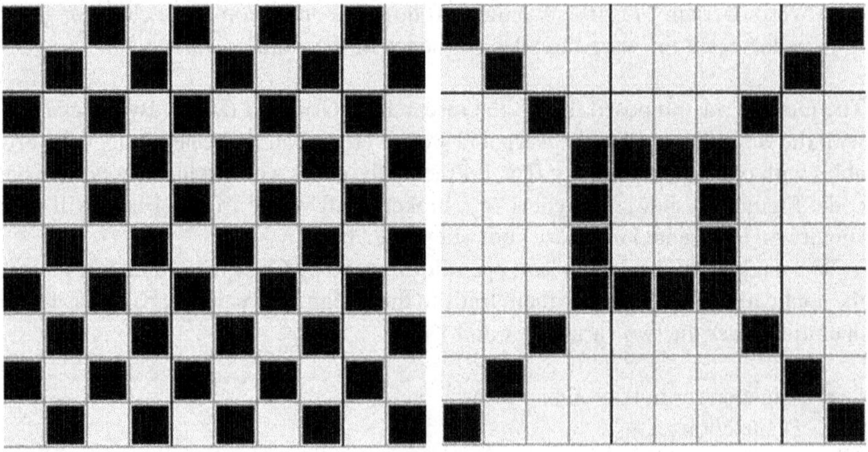

(a) (b)

FIGURE 8.2 Basic weave patterns: (a) pattern *A*—the plain weave and (b) pattern *B*—a motif.

8.2 THE BOOLEAN OPERATIONS IN WEAVE PATTERNS' DESIGN

The basic approach to the implementation of the Boolean operations requires the use of two matrices *A* and *B* with an equal number of rows *i* and columns *j*. In the case of the weave patterns, this means to select two basic weaves *A* and *B* with the same warp repeats $\left(R_{wpA} = R_{wpB}\right)$ and the same weft repeats $\left(R_{wfA} = R_{wfB}\right)$.

TABLE 8.1
Boolean Operations and Values of the Variables

Variable	Boolean Operation	Values of the Variables			
X		1	1	0	0
Y		1	0	1	0
$\sim X$	not (negation)	0	0	1	1
$\sim Y$	not (negation)	0	1	0	1
$X+Y$	or (disjunction)	1	1	1	0
$X \times Y$	and (conjunction)	1	0	0	0
$X \oplus Y$	exclusive or	0	1	1	0
$X \leftrightarrow Y$	equivalence	1	0	0	1
$X \rightarrow Y$	implication	1	0	1	1
$Y \rightarrow X$	implication	1	1	0	1

Let *A* and *B* be preliminary selected basic weave patterns, worked out in a 10 × 10 net. The first basic weave *A* is the plain weave (Figure 8.2a), chosen because of two reasons:

- It is the most straightforward and most frequently used pattern.
- It has an equal number of warp and weft overlaps.

The second basic weave *B* is a motif with a weft-faced texture (Figure 8.2b).

Figures 8.3 through 8.8 present the results of the application of the six Boolean operations over the selected basic weaves *A* and *B*. The only exception is the *negation*, which leads to the negative view of the basic patterns.

The analysis of the new patterns shows that the application of the Boolean operations can completely change the view of the basic weaves. Figure 8.3a shows the new weave pattern (one repeat) after use of the *disjunction* operation $(A+B)$: the operation breaks the cell-like effect of the motif *B* (Figure 8.2b), which results in a broken diagonal line. Figure 8.3b illustrates black warp/white weft threads color combination to show the cloth effect (three repeats in both warp and weft directions). Figure 8.3c is the weave view of the fabric, using the same color pattern as in Figure 8.3b.

The *conjunction* operation $A \times B$ transforms the plain weave to a new weave pattern (Figure 8.4a), similar to twill $\frac{1}{9}Z$ with a broken twill line, keeping the weft-faced view of the motif B. Figure 8.4b and c shows again the color pattern and the fabric view of the resulting weave.

On the contrary, the *exclusive or* operation $(A \oplus B)$ leads to a new weave with an equal number of warp and weft overlaps, transforming the weft-faced design of the motif *B*. Figure 8.5a–c shows the pattern of the designed weave, color view of 3 × 3 repeats, and the cloth appearance, respectively.

The *equivalence* operation $(A \leftrightarrow B)$ transforms the basic patterns to a new weave (Figure 8.6a), which is the negative version of the texture in Figure 8.4a. The result is logical, taking into account the relationship between the two operations, discussed in the next section. The color view of 3 × 3 repeats of the designed weave is presented in Figure 8.6b and the cloth view is presented in Figure 8.6c.

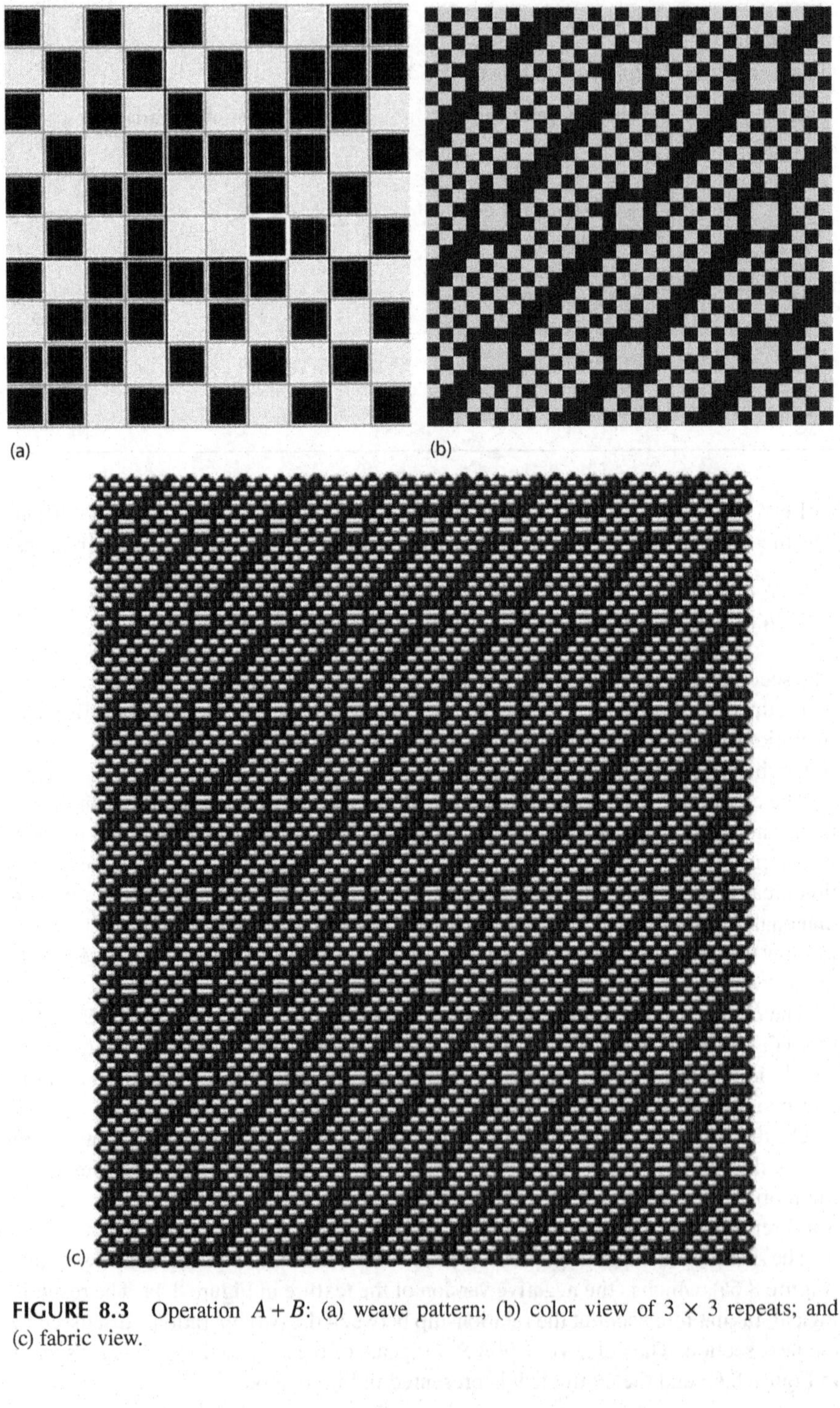

FIGURE 8.3 Operation $A+B$: (a) weave pattern; (b) color view of 3×3 repeats; and (c) fabric view.

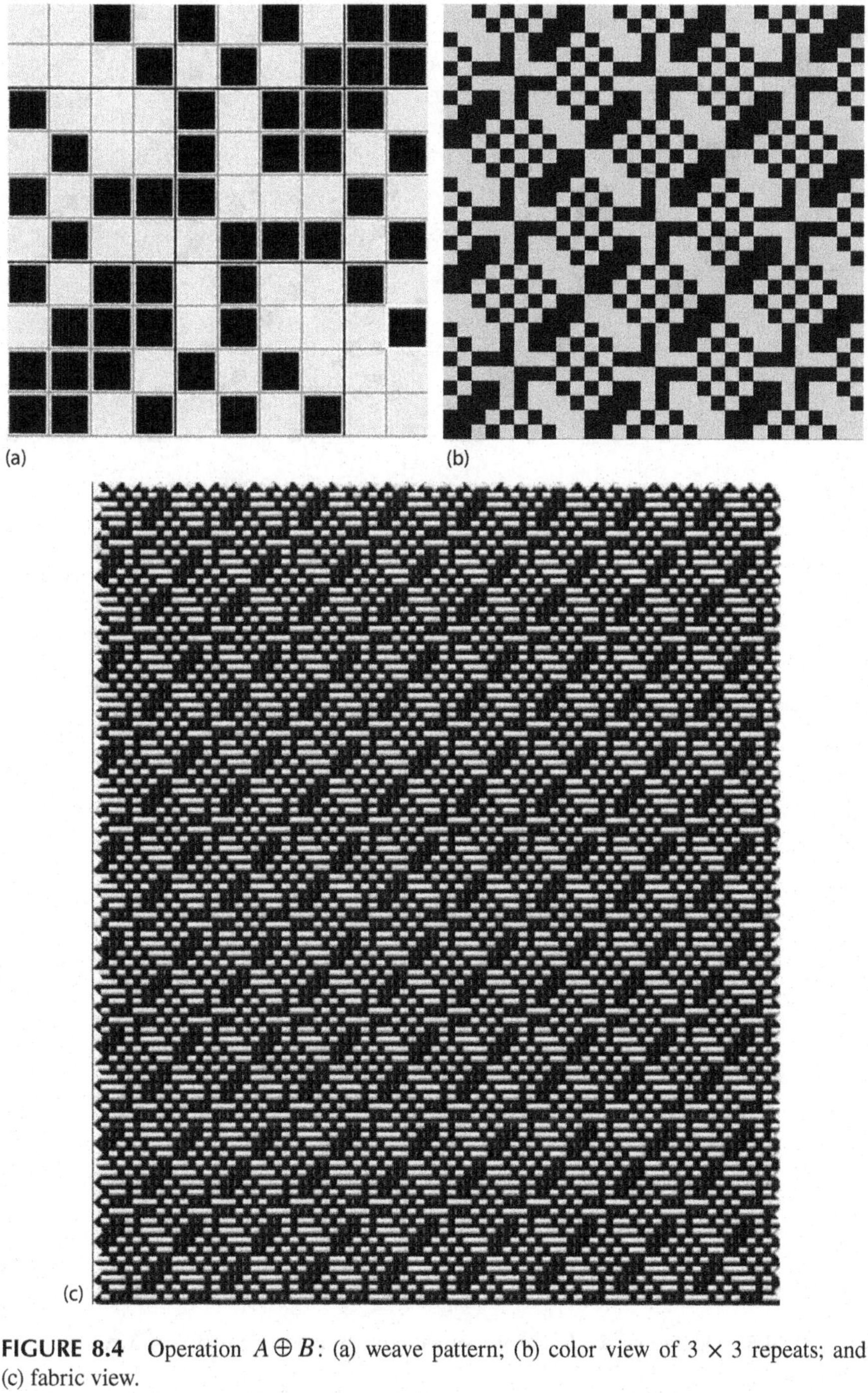

FIGURE 8.4 Operation $A \oplus B$: (a) weave pattern; (b) color view of 3 × 3 repeats; and (c) fabric view.

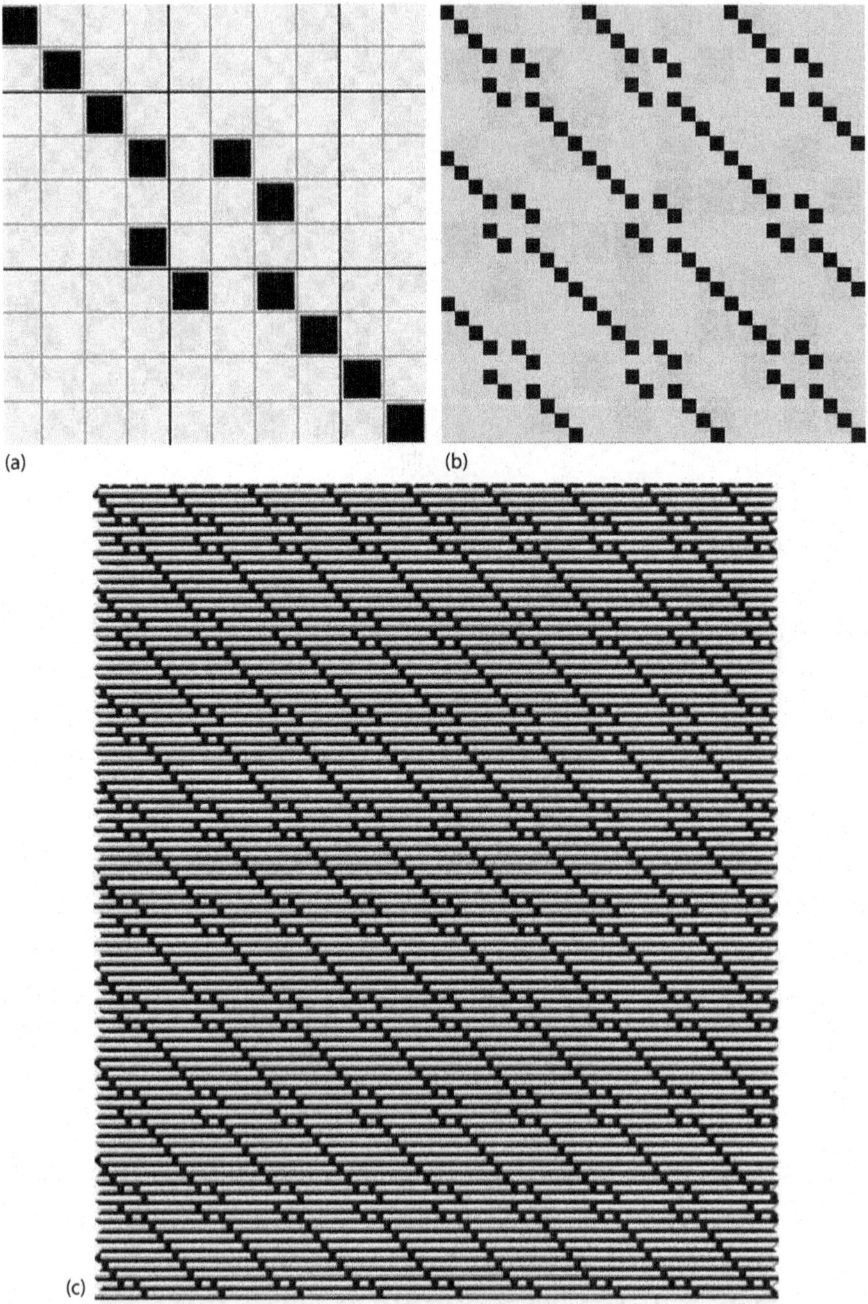

FIGURE 8.5 Operation $A \times B$: (a) weave pattern; (b) color view of 3 × 3 repeats; and (c) fabric view.

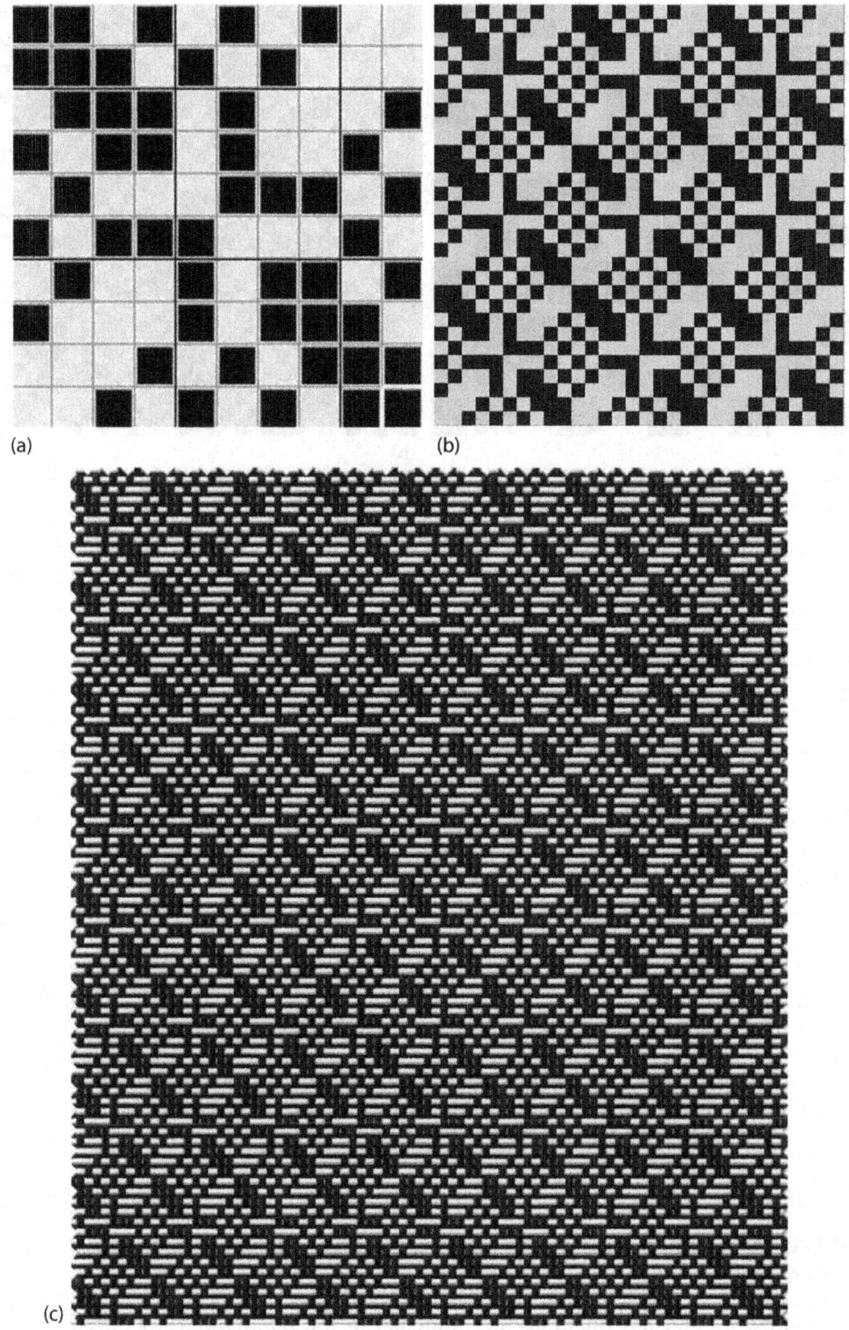

FIGURE 8.6 Operation $A \leftrightarrow B$: (a) weave pattern; (b) color view of 3 × 3 repeats; and (c) fabric view.

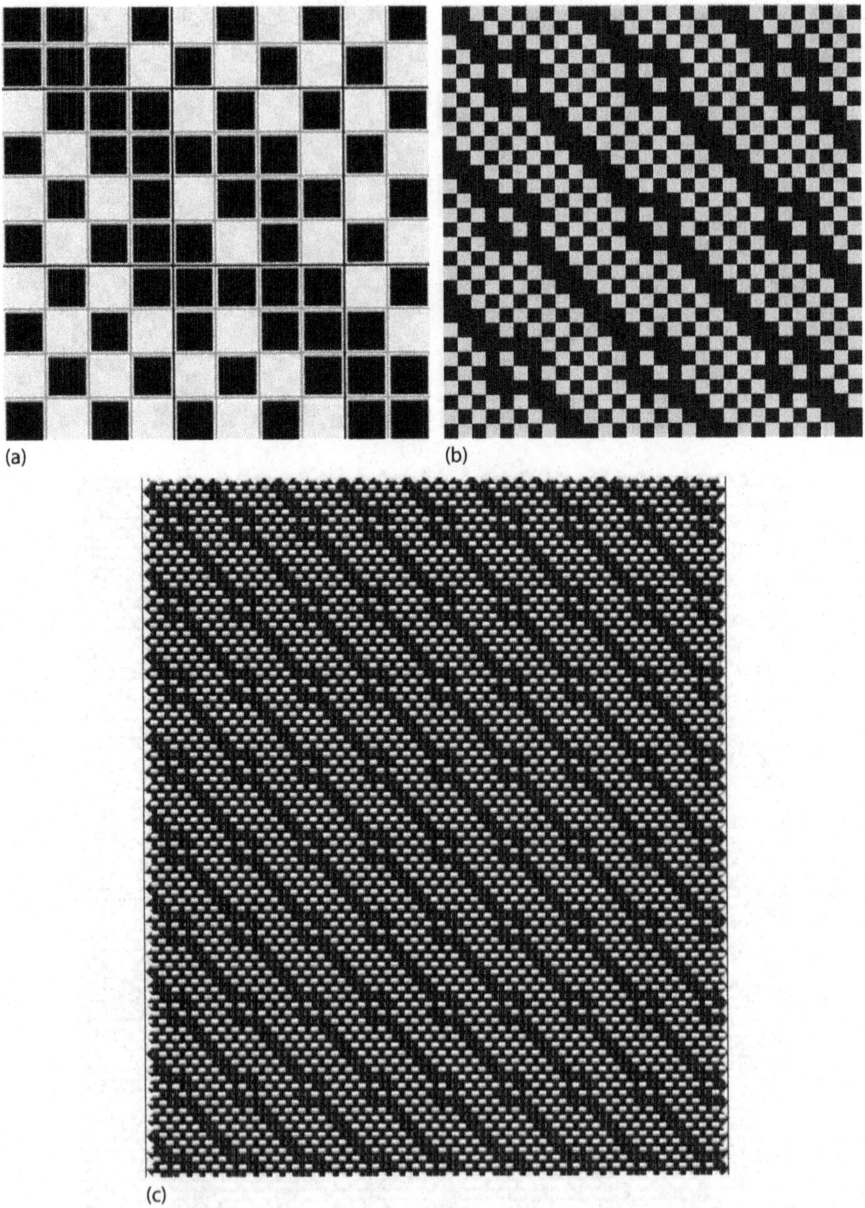

FIGURE 8.7 Operation $A \rightarrow B$: (a) weave repeat; (b) color view of 3×3 repeats; and (c) fabric view.

Figures 8.7 and 8.8 illustrate the influence of the *implication* operation on the basic weave patterns. Obviously, the result depends on the sequence of the variables A and B. The operation $A \rightarrow B$ leads to a new weave (Figure 8.7a), similar to the repeat in Figure 8.3a but still different (with left-handed broken twill diagonal). The operation $B \rightarrow A$ creates a new weave design (Figure 8.8), which is disparate

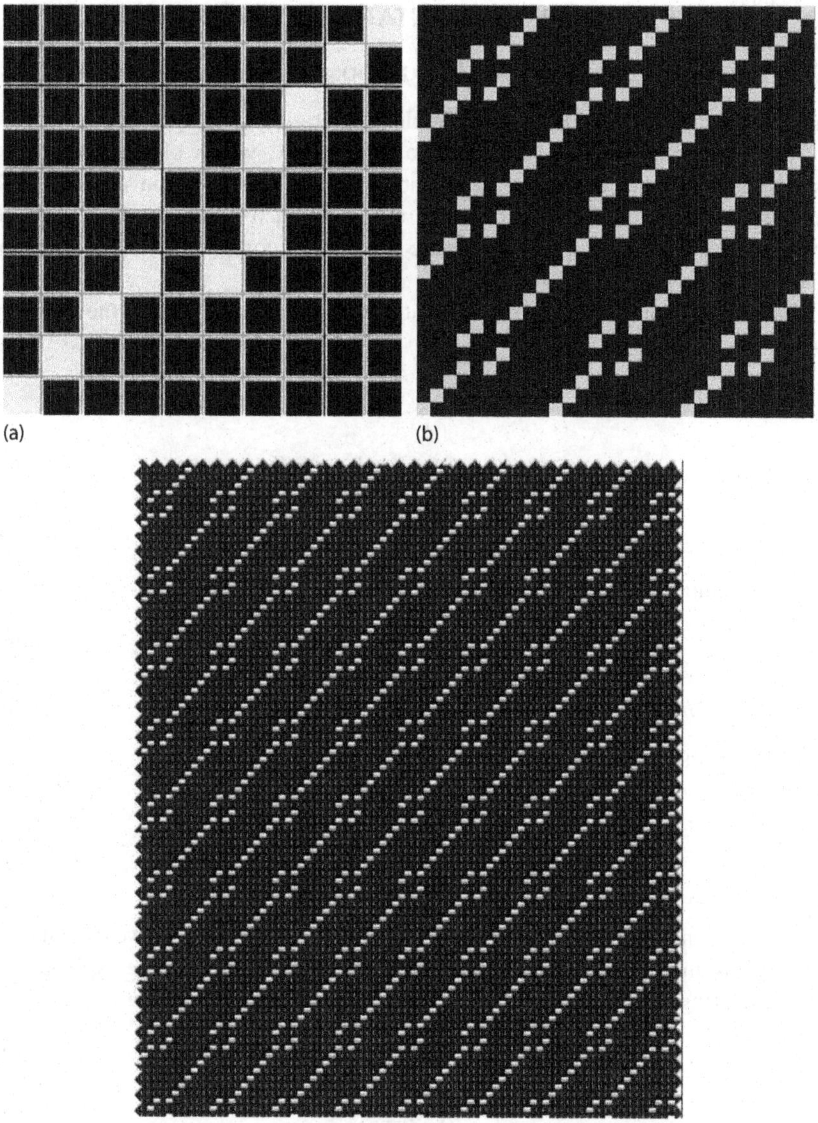

(a) (b) (c)

FIGURE 8.8 Operation $B \rightarrow A$: (a) weave repeat; (b) color view of 3 × 3 repeats; and (c) fabric view.

from Figure 8.7. The pattern repeat (Figure 8.8a) is with warp-faced texture, being both negative and mirror images of the weave repeat in Figure 8.5a.

The application of the Boolean operations does not require any special software or equipment. However, the method can lead to quicker results and faster design process if specialized software is used. Such software was developed and proposed by Kyosev and Angelova (2005).

8.3 FEATURES OF THE IMPLEMENTATION OF THE METHOD

8.3.1 Effect of the Basic Weave Designation

The result of the application of the Boolean operations, listed in Table 8.1, does not depend on the sequence of the basic weaves, i.e., which of the patterns will be denoted as *A* (and will have the variables *X*) or *B* (and will have variables *Y*). The only exception is the *implication* operation, in which the choice of the variables (or their sequence) is of importance for the result. Therefore, two different operations are considered: $X \rightarrow Y$ and $Y \rightarrow X$.

The Boolean operations over the matrices *A*, *B*, and *C* obtain the following features.

- Distributivity:

$$A\times\left(B+C\right)=A\times B+A\times C \tag{8.1}$$

$$A+\left(B\times C\right)=\left(A+B\right)\times\left(A+C\right) \tag{8.2}$$

- Commutativity:

$$A\times B=B\times A \tag{8.3}$$

$$A+B=B+A \tag{8.4}$$

- Associativity:

$$\left(A\times B\right)\times C=A\times\left(B\times C\right) \tag{8.5}$$

$$\left(A+B\right)+C=A+\left(B+C\right) \tag{8.6}$$

The application of the Boolean operations over two equal basic weaves (*A* and *A*) is meaningless and does not lead to obtaining a new weave design. The explanation is in the following dependences, which are valid in the Boolean algebra:

$$A+A=A \tag{8.7}$$

$$A\times A=A \tag{8.8}$$

$$A\oplus A=0 \tag{8.9}$$

$$A\leftrightarrow A=1 \tag{8.10}$$

$$A\rightarrow A=1 \tag{8.11}$$

The expression (8.9) means that all overlaps in the repeat of the new weave will be weft overlaps. The meaning of expressions (8.10) and (8.11) is the opposite: all overlaps in the repeat of the new weave will be warp overlaps. In both cases, there will not be a weaving process as the warp and weft sets will not interlace.

De Morgan's laws (Radojevic, 2006) have to be taken into account when a Boolean operation is applied over the basic weaves and their negative versions, namely:

$$\sim(A+B)=\sim A\times\sim B \tag{8.12}$$

$$\sim(A\times B)=\sim A+\sim B \tag{8.13}$$

8.3.2 Effect of the Basic Weave Commence

The treatment of the weave patterns as matrices means that every interlacement between a warp and a weft yarn is considered as a discrete element X_{ij} or Y_{ij} from the matrices (the basic weaves) $\|A\|$ and $\|B\|$. The progressive change of the places of the row i or column j does not lead to the new appearance of the fabric, except in the zones near the selvages. Such a progressive change of the places of the rows and columns, however, is very important when Boolean operations will be applied over the basic weaves.

To prove this conclusion, two basic weaves were taken: broken twill weave from twill $\frac{2\,.}{.\,2}Z$ (Figure 8.9a) as a basic weave A and weft rib $\frac{2\,.}{.\,2}$, developed in a net of 4 × 4 threads, as a basic weave B (Figure 8.9b). Two Boolean operations – *disjunction* $(X+Y)$ and *exclusive or* $(X\oplus Y)$ – were applied over the basic weaves. Four different commence of the first weave A were used, changing the consequence of the warp threads (Figure 8.10), while keeping the second weave B constant.

The results from the application of the *disjunction* operation are shown in Figure 8.11. Figure 8.12 illustrates the effect of the use of the *exclusive or* operation.

Obviously, the obtained results (the new weave repeats) are different depending on the different warp threads commence of weave A (Figures 8.11 and 8.12). This is a logical consequence of the change of the matrix A.

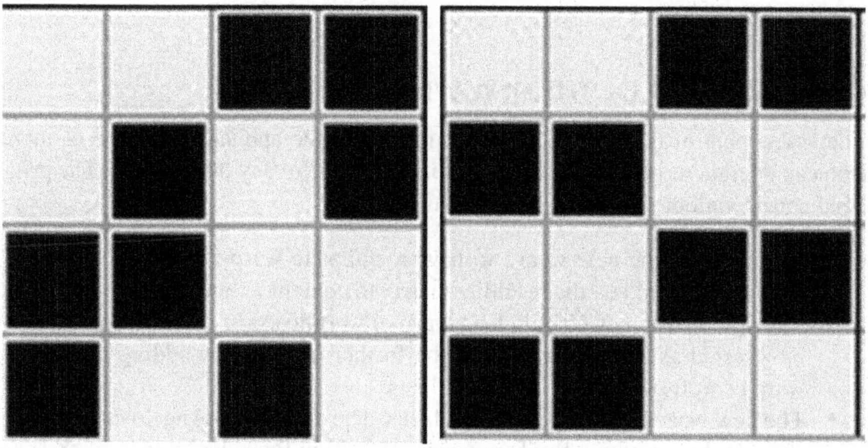

(a) (b)

FIGURE 8.9 Basic weaves in a 4 × 4 net: (a) weave A – broken twill weave from twill $\frac{2\,.}{.\,2}Z$ and (b) weave B – weft rib $\frac{2\,.}{.\,2}$.

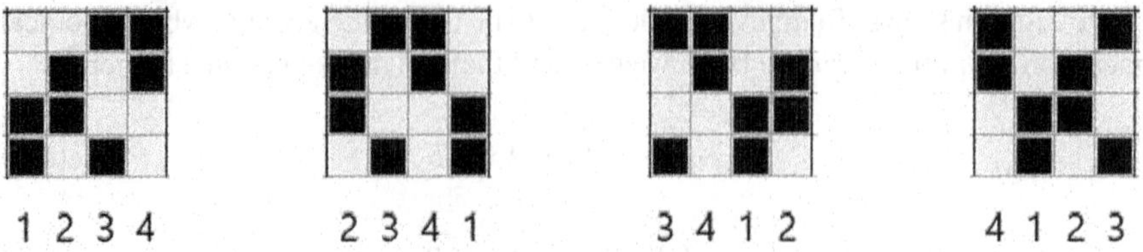

FIGURE 8.10 The four different warp threads commence of the weave A.

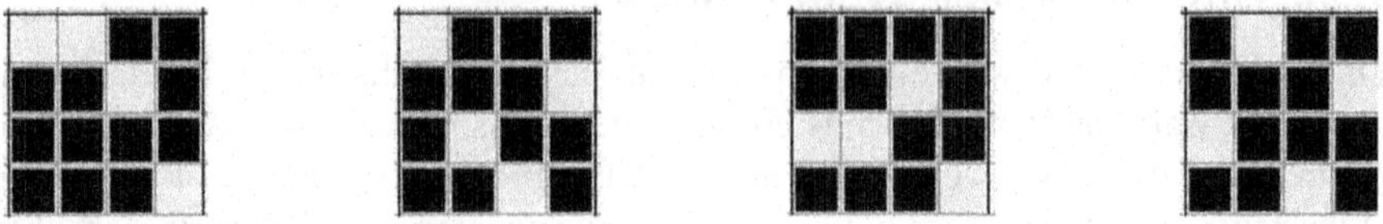

FIGURE 8.11 Results from the application of the *disjunction* operation.

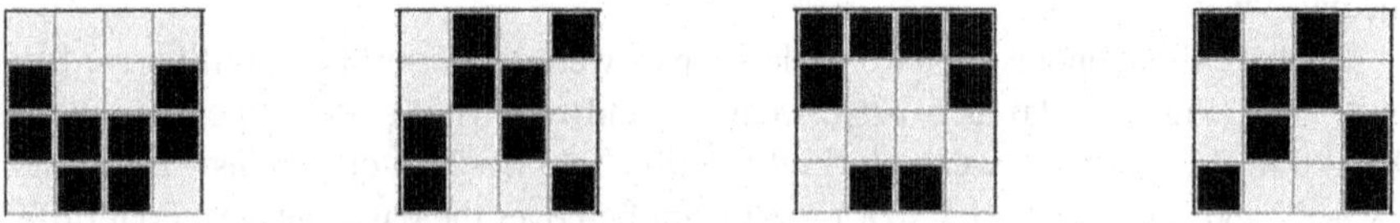

FIGURE 8.12 Results from the application of the *exclusive or* operation.

The analysis of the new weaves shows that the two applied Boolean operations lead to failed results in the cases when the commence of the pattern *A* starts with the first and third warp threads: there is an absence of interlacing between a weft thread and all warp threads within the repeat of the new weave. All other designs can be used for cloth weaving. They also can be used as a basis for other methods for creating new weaves (e.g., as motifs).

8.4 ASSESSMENT OF THE NEW WEAVE

The assessment of the new patterns obtained after the application of one or more Boolean operations does not differ from the evaluation of any other weave. The principal considerations can be summarized as follows:

- It is possible the new weave to have a complete warp-faced or complete weft-faced view, i.e., the resulting matrix to contain only 0 or only 1.
- The new weave may contain long warp or weft floats. In such a case, it can be rejected as a negative result or be further modified by adding random warp or weft overlaps to break the floats.
- The new weave can have different faces. The analysis in Angelova (2005) showed that only two Boolean operations (from those listed in Table 8.1) lead to the generation of a new weave patterns with balanced structure (with an approximately equal number of both warp and weft overlaps in the repeat). These are *exclusive or* $(X \oplus Y)$ and *equivalence* $(X \leftrightarrow Y)$.

- The new weave can be symmetrical or asymmetrical. Additional design techniques may be applied to it (rotation, mirror image, negative view, etc.) to obtain a composition based on two or more repeats of the new weave.

8.5 DESIGN EXPERIMENTS

To demonstrate the results from the application of the Boolean operations over a set of two basic weaves, two types of weaves were chosen:

- a constant basic weave: weft-faced sateen $\frac{1.}{.5}Z^2$ (Figure 8.13a);
- variable weaves:
  - twill weave $\frac{1.}{.4}Z$ with a particular commence of the warp threads (Figure 8.13b);
  - twill weave $\frac{1.}{.4}S$ (Figure 8.13c);
  - hopsack weave $\frac{.3}{2.}$ (Figure 8.13d);
  - a motif with a cell-like effect (Figure 8.13e).

Figure 8.14 summarizes the results from the application of the Boolean operations over each pair of basic weaves: a total of 24 new patterns (the *negation* operation was neglected, as it only changes the view of the basic weave). The appropriateness of the new designs for the weaving of cloth is commented from the point of view of the resulting interlacing between the warp and weft threads and marked with *yes/no* above each weave repeat.

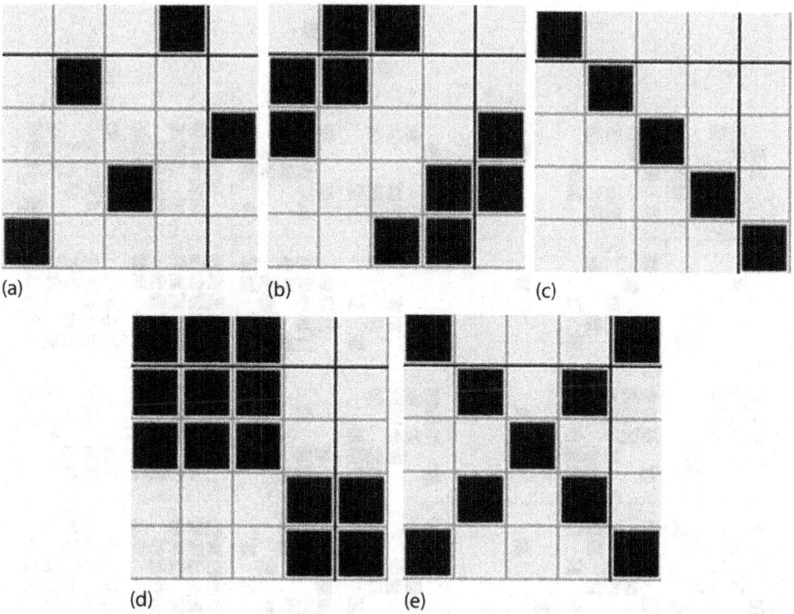

FIGURE 8.13 Application of the Boolean operations: (a) sateen $\frac{1.}{.5}Z^2$, a constant basic weave; (b) twill $\frac{1.}{.4}Z$, a variable basic weave; (c) twill $\frac{1.}{.4}S$, a variable basic weave; (d) hopsack $\frac{.3}{2.}$, a variable basic weave; and (e) a motif, a variable basic weave.

Having randomly selected basic weaves, we can conclude that 13 of the new designs are suitable for the production of woven cloth and 11 of them have warp or weft threads (at least one in the repeat) that do not interlace with the opposite set of yarns. These 11 designs can be additionally manipulated to break the yarn float. The *conjunction* operation did not lead to the generation of a quality weave repeat in none of the cases. However, in other combination of basic weaves, it can lead to positive results.

Some of the patterns in Figure 8.14 are used to show the great opportunities, given by the new weaves, for creating a variety of new weave designs.

The weave pattern 4 (Figure 8.14) is developed further as a composition, in which the designed pattern rotates clockwise to 90°, thus forming a weave with a square repeat $\left(R_{wp} = R_{wf} = 10\right)$. Figure 8.15a illustrates the repeat of the composition, and the view of the fabric is shown in Figure 8.15b.

The additional design over the weave pattern 18 (Figure 8.14) involves a composition in a 10 × 10 net. The composition includes a horizontal mirror view of the weave pattern and a vertical mirror view of the pattern and the first mirror view. The resulting repeat is shown in Figure 8.16a and the fabric view is demonstrated in Figure 8.16b.

Two of the new weave patterns from Figures 8.14 through 8.16 were applied further to make a composition in a chessboard order. The repeat of the pattern $\left(R_{wp} = R_{wf} = 10\right)$ is shown in Figure 8.17a and the fabric view is presented in Figure 8.17b.

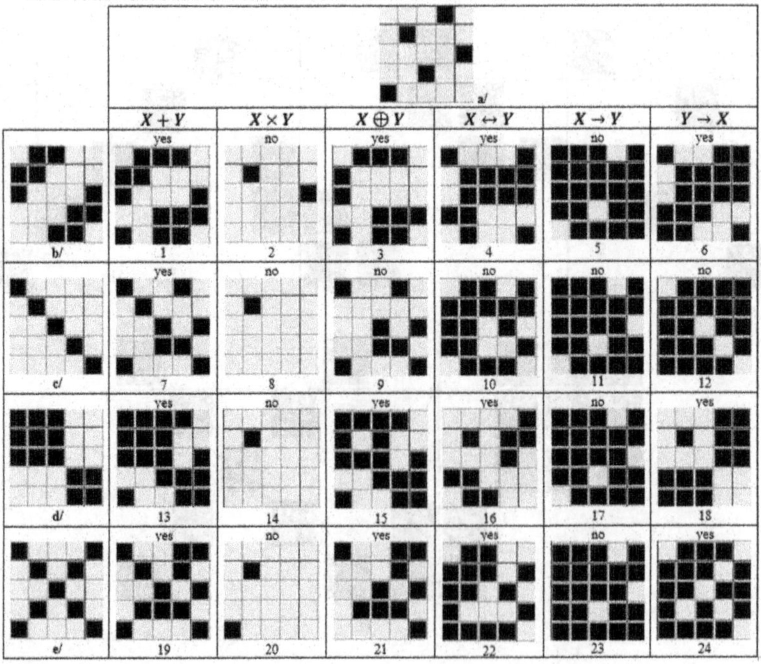

FIGURE 8.14 Results from the application of the Boolean operations over four pairs of basic weaves.

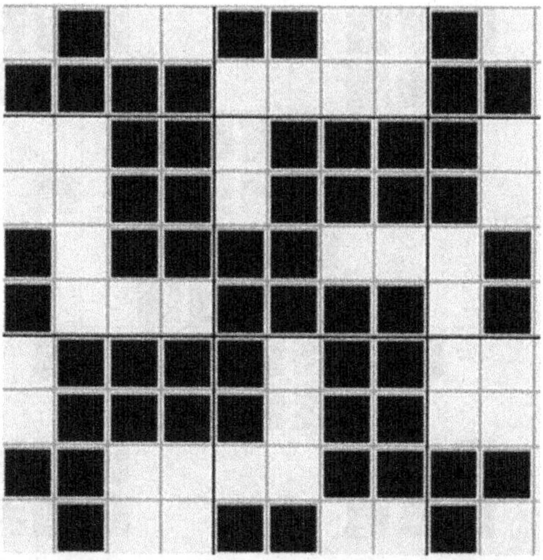

(a)

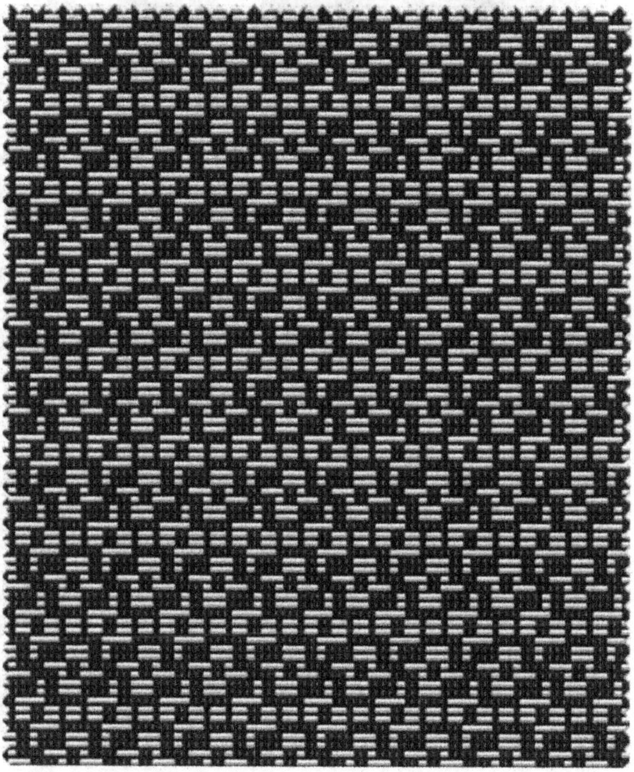

(b)

FIGURE 8.15 A composition with the new weave pattern 4 (Figure 8.14): (a) repeat of the composition and (b) fabric view.

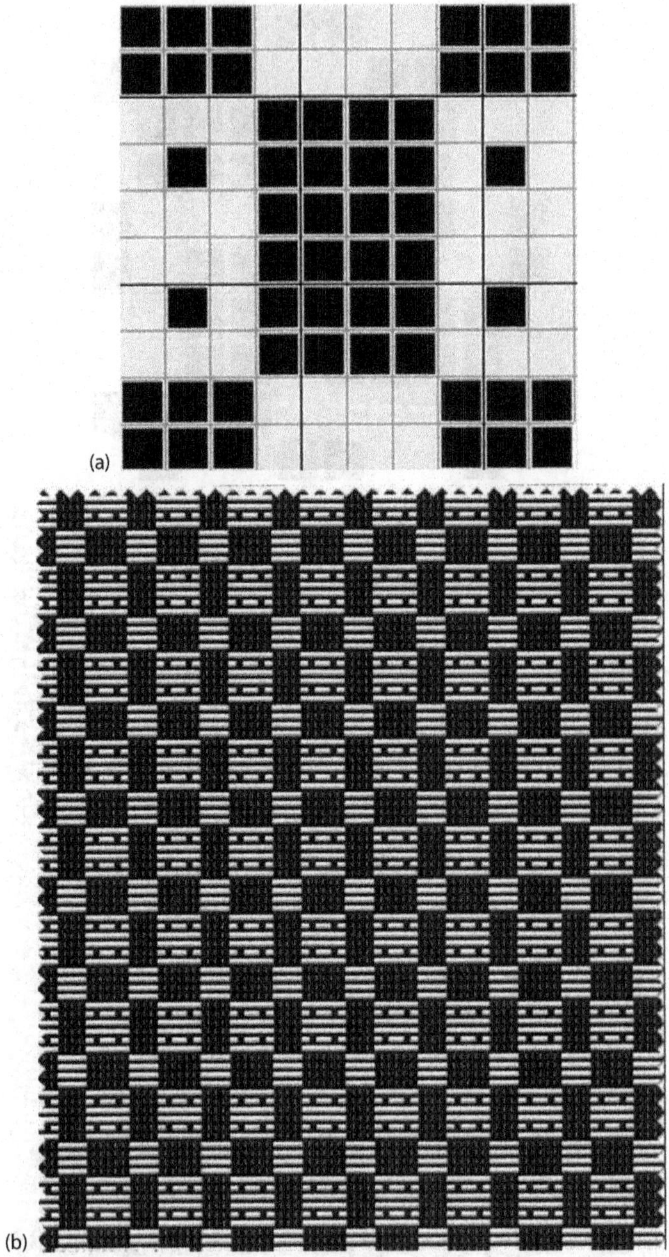

FIGURE 8.16 A composition with the new weave pattern 18 (Figure 8.14): (a) repeat of the composition and (b) fabric view.

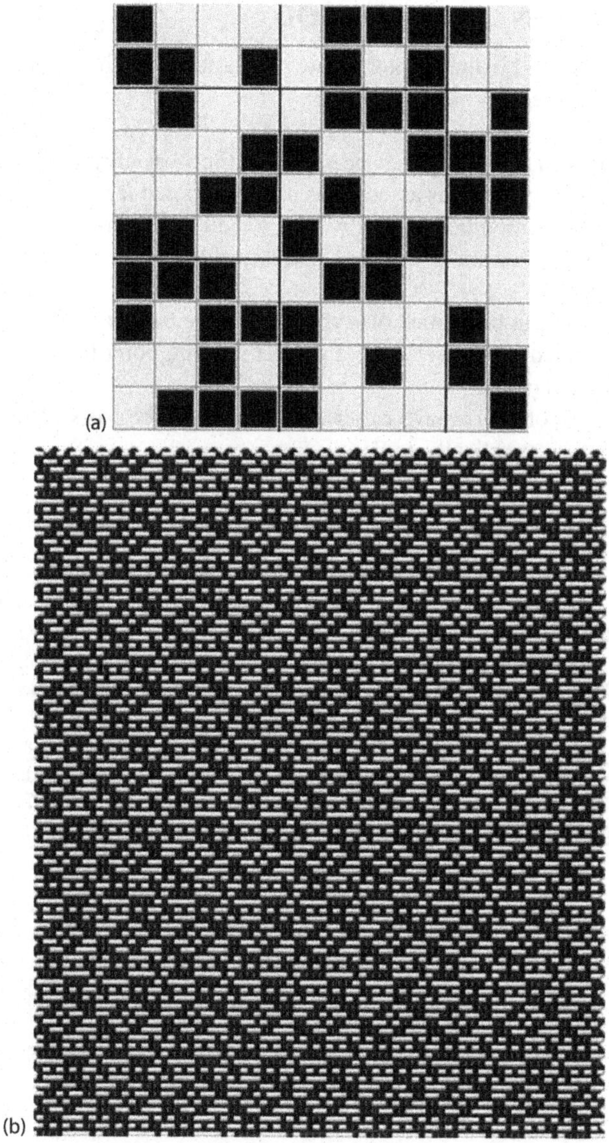

FIGURE 8.17 A chessboard composition with the new weave patterns 15 and 16 (Figure 8.14): (a) repeat of the composition and (b) fabric view.

8.6 APPLICATION OF THE METHOD

The method discussed in the chapter allows different ways to implement the Boolean operations in the creation of new weaves:

- The generated new weave may be applied entirely or only part of it to be used.
- In case the new weave is assessed as unappropriated for a fabric weaving, it can be further modified manually (e.g., to add new overlaps).
- Two or more operations can be consecutively applied over a pair of basic weaves.
- The method can be applied over three and more basic weaves.
- The generated new weaves are a perfect starting point for composition of new weave repeats.
- The variety of new weave patterns, obtained on the basis of the Boolean operations, is practically endless.

REFERENCES

Griswold, R. E. (2002). Designing weave structures using Boolean operations, Part 1, Available from http://www.cs.arizona.edu/patterns/weaving/webdocs/gre_bol1.pdf.

Angelova, R. A. (2005). Designing new weaving patterns using Boolean operations. *Proceedings from X International Conference of the Faculty of Power Engineering, EMF'2005* (Vol. 2, pp. 137–142), Sozopol, Bulgaria.

Kyosev, J., & Angelova, R. A. (2005). Specialized software for generation and investigation of the woven pattern. *Proceedings from Aachen Textile Conference*, Aachen, Germany.

Radojevic, D. (2006). Interpolative realization of Boolean algebra. In *2006 8th Seminar on Neural Network Applications in Electrical Engineering* (pp. 201–206). IEEE.

9 New Weave Patterns Based on Musical Scores

9.1 THE CONCEPT OF THE METHOD

The background of the idea lies in the similarity between the notation and the 2D coding of weave patterns. Such a design method has not been presented in the literature, except in a publication by the present author (Angelova, 2017).

The method uses the musical pieces and their notations as inspiration for the creation of new weave patterns. The method does not state that the textile design would "reflect" the musical piece, its temper, and suggestion. It is a pure "technical" relationship.

However, connotation between the color design and the music can be found by using proper colors, color alternation, and combination.

9.2 DESIGN METHOD

9.2.1 Specifics of the Weave Patterns

The application of a binary code to display the weave patterns requires a weaving diagram (a net), where it is considered that the warp threads are between two vertical lines, while the weft threads are assumed to be located between two horizontal lines (Figure 9.1a).

Even the cursory examination of the notation in music shows a similar analogy: some of the notes are placed on the lines of the staff, and some of them between the lines (Figure 9.1b). In C-major scale, the notes re (D), fa (F), la (A), do (C), mi (E), and sol (G) are located between the lines (Figure 9.2a), and the rest on the lines of the staff (Figure 9.2b). In certain circumstances, the encoding of a musical piece can be converted into the encoding of a weave pattern. A few bars only are sufficient to create a new, unique weave that is not subject to copyright or license issues.

9.2.2 Transformation Method

The transformation method concerns the transition from musical notation to a weave diagram. Several rules, presented shortly, were defined and refined to overcome the lack of correspondence between the basic principles in notation and 2D weave patterns encoding.

The most significant discrepancy is probably that half of the musical signs occupy the lines of the staff, while the weave diagram can use only the spaces (squares) between the lines. When transforming the notation of a musical piece into a weave diagram, every note on the staff is considered as a warp overlay. Each subsequent note is considered to be a warp overlay of the next thread.

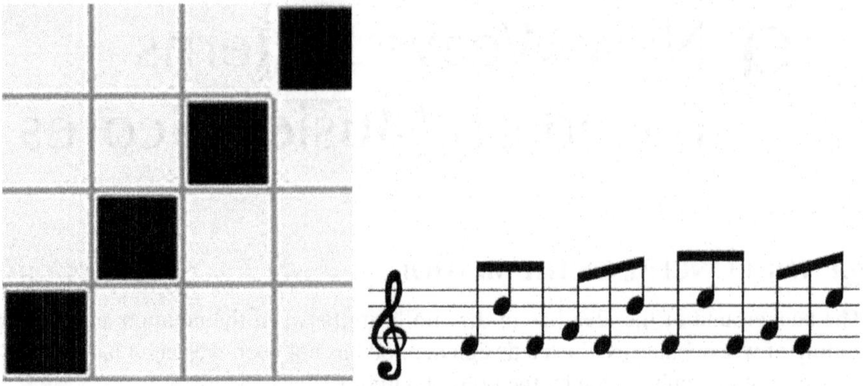

(a) (b)

FIGURE 9.1 The analogy between a weave pattern and a musical score: (a) twill weave $\frac{1}{3}Z$ and (b) a part of a musical score.

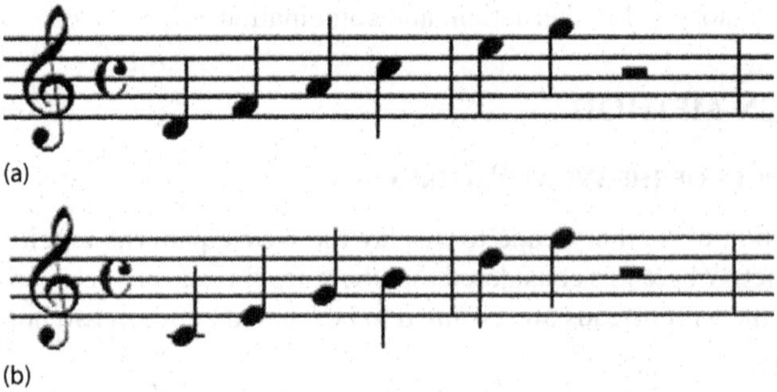

FIGURE 9.2 Note signs: (a) written between the lines of the staff and (b) written on the lines of the staff.

Special attention is paid on cords: they are transformed to two or more warp overlays of the same warp thread.

The notes, written between the lines of the staff, are considered as warp overlays, marked directly in the weave diagram (Figure 9.3a). After several attempts for a different solution, the notes, written on the lines of the staff, are considered as warp overlays in the row above (Figure 9.3b) or the row below (Figure 9.3c).

The transformation method ignores the accidentals (flats or sharps); only the position of the note on the staff is considered to be important. The duration of the notes is also not taken into account for the moment, though it has the potential to be used as a source of additional transformation.

Figure 9.3 shows that the transformation from notes on the staff to a weave pattern is one and the same for different notes. For example, in the weave diagram, the sol note (G note) has the same position as la note (A note) if it is moved up, or as fa note (F note) if it is moved down. G-flat and G-sharp have the same position as G.

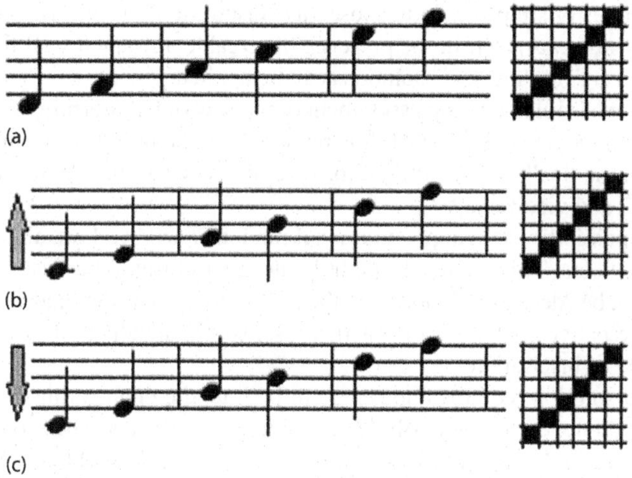

FIGURE 9.3 From notes to pattern transformation: (a) direct; (b) moving the notes from the line to the row up; and (c) moving the notes from the line to the row down.

As a result, several different musical notes, namely, sol (G), la (A), G-flat, G-sharp, A-flat, and A-sharp have the same position in the weave diagram. At first glance, this fact might provoke confusion, but it creates freedom in the transformation, which is necessary during the assessment stage of the generated new pattern.

9.2.3 Warp and Weft Repeats

The questions where to stop the transformation from the musical score to the weave pattern is very logical, as even the monotonous child song can have dozens of notes.

It can be seen that the musical staff itself limits the repeat in weft direction: perhaps it can reach 15 or 20 threads but in rare cases. The same is not valid for the repeat in warp direction: it can be as big, as long the musical piece is, but just theoretically. Practically, the shedding mechanism of the available weaving machine limits the repeat of the warp threads, and the designer has to take it into account.

In fact, there is no limitation which part of the musical piece can be taken into consideration for the transformation: it is just the inspiring precondition. The samples, presented in the study, show different approaches in this sense: from the use of the almost whole notation of a short, popular song, to the transformation of a couple of times from a complicated classical piece.

9.3 ASSESSMENT OF THE NEW WEAVE

The evaluation criteria of the resulting new weave pattern are the same as for the evaluation of the patterns produced by any other method. Particular attention should be paid to the appearance of the fabric and the yarns floats.

In most cases, the generated "musical" weave pattern would have a weft appearance. There is a risk of getting long weft floats that would make the fabric unstable.

If a monotonous or simple melody is used as "inspiration", a weft thread may never interlace with the warp set. In the presence of sizeable unwanted weft floats, different "engineering" or "musical" solutions can be applied.

The first one is related to the assessment of the way of transforming the notes that are located on the lines of the staff. It is possible that the used transformation of the notes "down the line" to be disturbed for several notes and these notes to be moved "up the line" and vice versa.

Another possibility is to apply additional random warp overlays, aimed at breaking the weft floats. This approach might be done carefully; otherwise, the original "melody" can be "destroyed". However, the most important is the final goal: to create a new weave pattern, which is of practical use; so, the notation could be used as a vague source of inspiration.

The third solution requires knowledge in both textile engineering and music, as it is a musical one: cords can easily break the large weft floats. Therefore, the transformation of a single note in a chord, which is in tune with the basic tone, requires additional warp overlays to be marked in the weave diagram. In musical pieces for piano, both right- and left-hand notations can be utilized as well; thus, the appearance of a chord to be of use for the pattern and the stability of the future woven fabric.

9.4 DESIGN EXPERIMENTS

The notations of four famous musical pieces were used to create new, *musical weave patterns*, and fabric versions of them (black warp and white weft threads), to demonstrate the unlimited possibilities of the developed method. The musical pieces were randomly chosen: the most famous song in the world, a popular jazz melody, and two classical compositions. Any piece of music can be applied, but the designed weave pattern has to be assessed after the transformation for production appropriateness. Possible color designs of the designed weave patterns can be seen in Angelova (2017).

9.4.1 Weave Pattern "Happy Birthday to You"

According to Guinness World Records (1997), "Happy Birthday to You" is the most popular song in the English language. Figure 9.4 presents the notation of the song, used for the weave pattern transformation.

A modification of the melody was applied to break the weft floats: instead of a single note mi (*E*), a chord of three notes was used in the first, third, and fifth bars. The derived *musical weave pattern* is shown in Figure 9.5a: its repeat is $R_{wp} = 25, R_{wt} = 5$. The largest weft float is over 11 warp threads (see the first weft thread), which is acceptable, especially when the warp yarns are with small linear density. If not, an additional one or more warp overlaps may be added. Figure 9.5b shows the cloth view.

9.4.2 Weave Pattern "Summer Time"

The notation of the piece "Summer Time" is presented in Figure 9.6. There is a repeat in the melody: the first four bars repeat from the ninth bar. Thus, the bars from

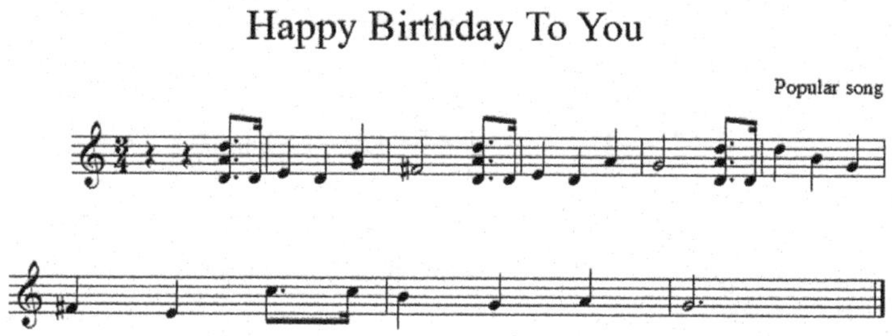

FIGURE 9.4 Notation of the song "Happy Birthday to You" with the changes, used for deriving the weave pattern.

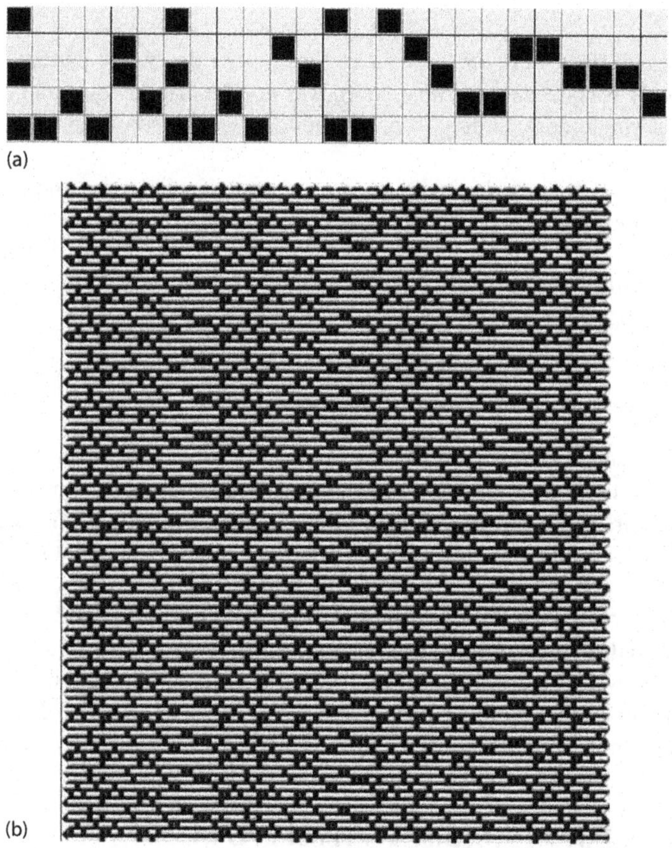

FIGURE 9.5 "Happy Birthday to You" musical weave pattern: (a) repeat and (b) fabric view.

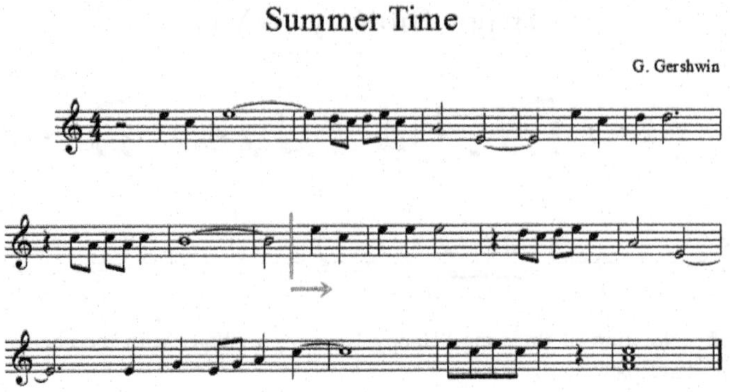

FIGURE 9.6 Notation of "Summer Time" with the changes in the final chord, used for deriving the weave pattern.

the ninth bar further were only considered to shorten the weave pattern. A change in the melody was needed: the final chord was not the original one, and it replaced the original single note, aiming to break the long weft float of the first three weft threads.

The repeat of the *musical weave pattern*, transformed from the notation, is shown in Figure 9.7a $\left(R_{wp} = 26, R_{wt} = 4\right)$. The largest weft float is over 11 warp threads (see the first and the fourth weft threads in the repeat), which can be broken with additional warp overlay, if needed. The view of the fabric is shown in Figure 9.7b.

9.4.3 Weave Pattern "The Moonlight Sonata"

The musical score of the Beethoven's one of the most famous compositions is too big to be used for a whole weave pattern. However, just a small part of it, even some bars only, can be an inspiration for the development of a new weave pattern. The notation for piano of the first two bars of the sonata is shown in Figure 9.8.

Only the very first bar (for right and left hand) was used to derive the weave repeat (Figure 9.9a); thus, the warp and weft repeats are $R_{wp} = 12, R_{wt} = 4$. The largest weft float is again over 11 warp threads (see the first weft thread in the repeat). The float may be broken with additional warp overlays if needed. The view of the fabric is demonstrated in Figure 9.9b.

9.4.4 Weave Pattern "Toccata and Fugue in D Minor"

Similarly to the example with "The Moonlight Sonata", only the first bar of the notation of "Toccata and Fugue in D Minor", shown in Figure 9.10, was considered to transform the musical score to a weave pattern. Moreover, only the first 10 notes (till the second pause) were taken to make a repeat of 10 warp threads

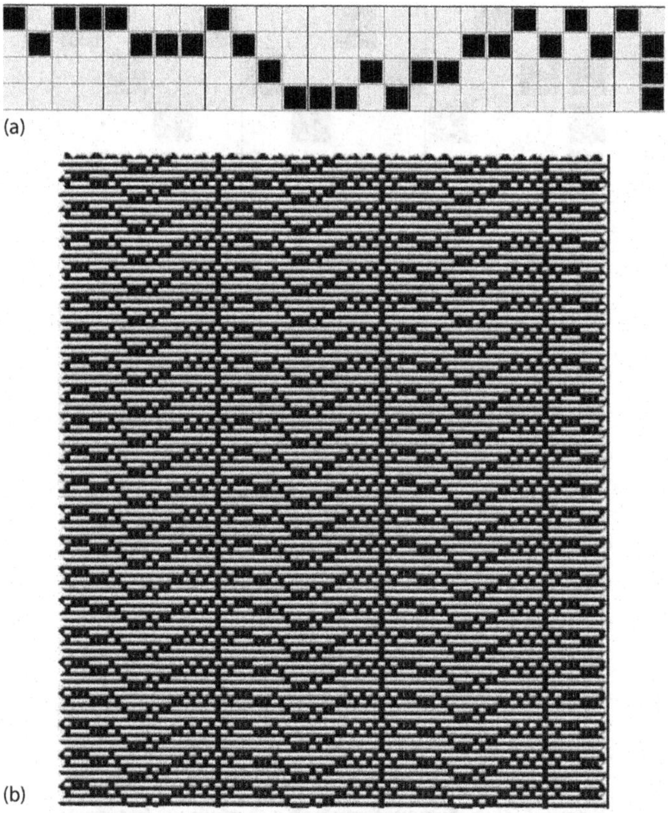

FIGURE 9.7 "Summer Time" musical weave pattern: (a) repeat and (b) fabric view.

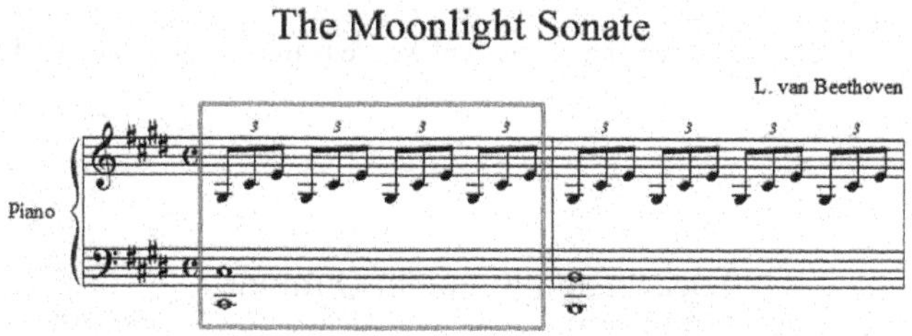

FIGURE 9.8 Notation of "The Moonlight Sonata" (first two bars).

$(R_{wp} = 10, R_{wt} = 7)$. The repeat itself is shown in Figure 9.11a, together with the fabric view (Figure 9.11b).

It has to be mentioned that the polyphonic melodies are very suitable for creation of weave patterns, as the risk for appearance of long weft or warp floats is smaller.

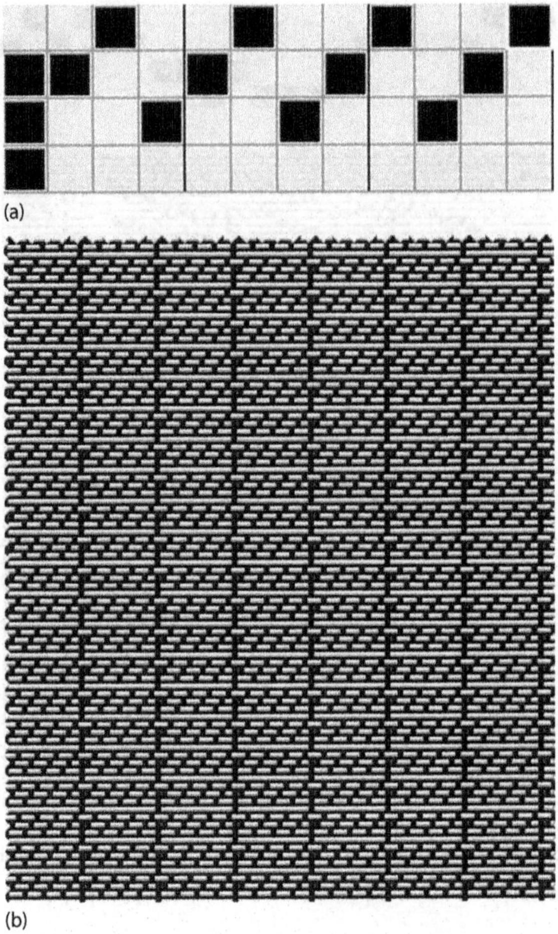

FIGURE 9.9 "The Moonlight Sonata" musical weave pattern: (a) repeat and (b) fabric view.

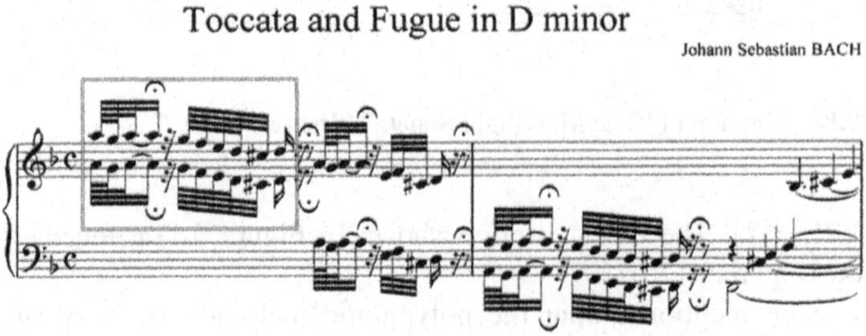

FIGURE 9.10 Notation of "Toccata and Fugue in D Minor" (first two bars).

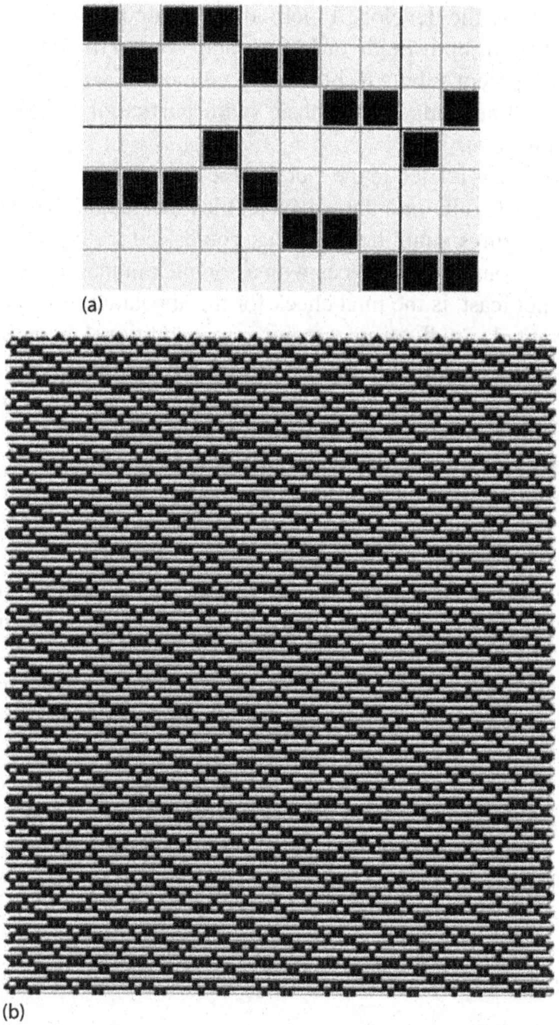

FIGURE 9.11 "Toccata and Fugue in D Minor" musical weave pattern: (a) repeat and (b) fabric view.

9.5 APPLICATION OF THE METHOD

The developed method and the four new weave designs, based on musical pieces could be considered as successful first attempts that open up additional opportunities for the design and production of original, unique fabrics. Having in mind the methods presented in Section I (Chapters 4 through 7), used up to now for the development of new weave patterns, it can be concluded that the novel method broadens the existing possibilities and may open a new field: the creation of cloth weaves based on similarity with art.

The limitations of the developed method are related mainly with the requirements for expertise in both fields of human activity: textile engineering (though the pattern development seems to be closely connected with art, it requires specialized technical knowledge) and music (with particular emphasis on notation readings and solfeggio).

Another limitation is related to the fact that the complexity of the musical notation cannot be "translated" fully into the woven textiles' language. The transformation to a weave pattern requires simplifications that could lead to the creation of not very attractive patterns from musical pieces with a complex notation.

The last, but not least, is the final check for the appearance of the pattern and the presence of unwanted yarn floats may require more additional work than the transformation from notation to a pattern itself. However, the possibility to make significant design changes in the "derived" weave pattern, inspired by a musical piece, leads to the final aim of the method: the creation of a new, unique pattern. In this sense, the method can be also used for initial inspiration only.

The weave pattern, based on a musical piece, can be further "worked", using other methods (e.g., mathematical methods of transformation) to create visually attractive new patterns and cloths.

Owing to the technical advantages, the method possesses marketing potential for the producers of woven textiles and designers. As a work in progress, the method will evolve from both theoretical point of view and practical application. The main direction of research is toward adequate use of the complex musical notation in the weave pattern, finding potential in the transformation of the duration of the notes.

ACKNOWLEDGMENT

Part of this chapter is published with the permission of https://www.tandfonline.com/ for the use of the manuscript (Angelova, 2017).

REFERENCES

Angelova, R. A. (2017). Design of weave patterns: when engineering textiles meets music. *The Journal of the Textile Institute*, *108*(5), 870–876.

Guinness World Records. (1997). *Guinness Book of World Records*. New York: Viking Press.

10 New Weave Patterns Based on Text Interpretations

10.1 THE CONCEPT OF THE METHOD

The mystical relationship between numbers and letters has been studied since ages in divinatory arts like numerology and cabala. The modern systems of transformation from letters to digits are usually based on the consequence of the alphabet letters, assigning them the digits from 1 to 9 repeatedly. Thus, the letter A corresponds to 1, B to 2, N to 5, R to 8, etc.

Once having a transformation matrix, every single word, a phrase, a sentence, or a longer text can be easily converted to a set of digits from 1 to 9.

10.2 TRANSFORMATION APPROACHES

10.2.1 Basic Approach

Table 10.1 illustrates the most used matrix (table) for the transformation of the English language alphabet from letters to digits from 1 to 9.

Let's imagine that a weave repeat involves nine weft threads and an unknown number of warp threads. Then, Table 10.1 can be used to place a warp overlap over each consecutive warp thread, starting from the first warp yarn in the repeat. Every single letter takes its place from the first to the ninth weft thread, following Table 10.1 and from the first warp thread to the last, following the consequence of the letters in the words.

Figure 10.1 is used to illustrate the transformation of the famous Shakespeare's phrase "To be or not to be" to a weave pattern. The warp repeat of the pattern is 13, as the number of the letters in the phrase is 13. The weft repeat is 9.

The location of the letters on the weave net is shown in Figure 10.1a, thus allowing a better understanding of the relationship between the letters and the weave pattern. Figure 10.1b illustrates the transformation to a weave pattern and Figure 10.1c illustrates the view of the fabric.

The result in Figure 10.1b is not a weave repeat that can be used for the production of a fabric: there are five weft threads in the repeat that do not interlace with any of the warp threads. The reason is that the phrase "To be or not to be" consists of similar letters and is quite short to involve all digits from 1 to 9.

TABLE 10.1
Transformation of the English Language Alphabet to Numbers from 1 to 9

Digits	Letters		
1	A	J	S
2	B	K	T
3	C	L	U
4	D	M	V
5	E	N	W
6	F	O	X
7	G	P	Y
8	H	Q	Z
9	I	R	

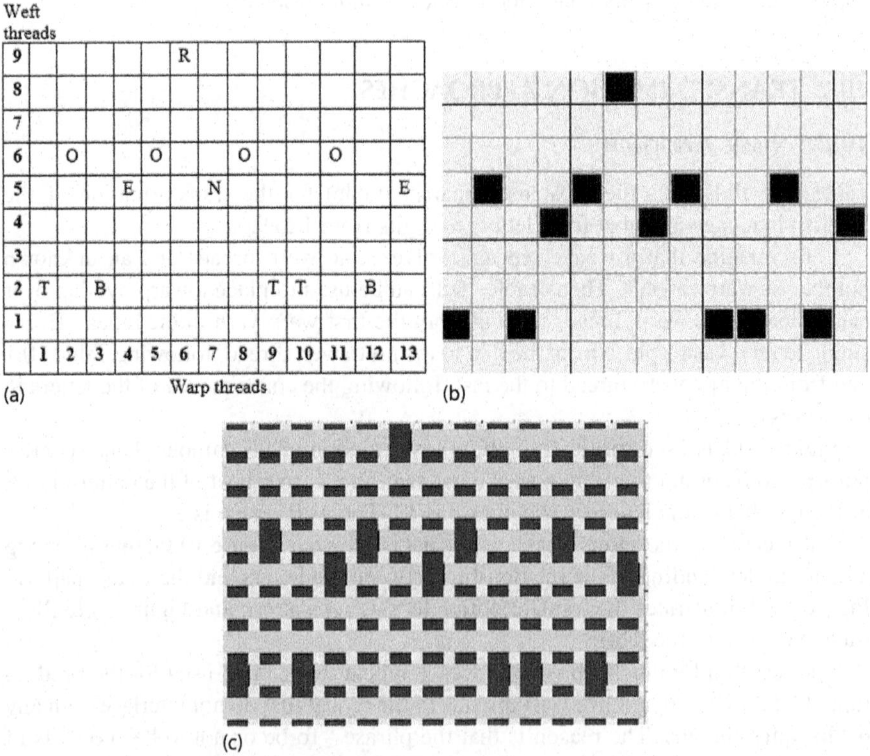

FIGURE 10.1 The transformation of the Shakespeare's phrase "To be or not to be" to a weave pattern with $R_{wf} = 9$: (a) the location of the letters in the weave net; (b) the weave pattern; and (c) fabric view.

10.2.2 Application of Different Transformation Matrices

One way to avoid the problem with missing warp and weft overlaps is to decrease the size of the transformation matrix, using a shorter set of digits. Table 10.2 proposes a matrix, based on the digits from 1 to 4. Thus, much more letters correspond to the same digit, and the problem of empty rows in the weave pattern may be avoided.

Based on the transformation in Table 10.2, a new weave repeat is developed for the phrase "To be or not to be". The result is shown in Figure 10.2, where the location of the letters in the weave net (Figure 10.2a), the designed weave pattern (Figure 10.2b), and view of the cloth (Figure 10.2c) are consecutively presented.

The transformation, based on Table 10.2, is much more appropriate, compared to the transformation, based on Table 10.1. There are no longer floats in the weave repeat (Figure 10.2b), and the view of the fabric (Figure 10.2c) shows attractive weave design with random zig-zag effect.

A logical question here is why the weft repeat is chosen to be $R_{wf} = 4$, isn't it much more appropriate to restrict the warp repeat to $R_{wp} = 4$? Indeed, the smaller warp repeat is beneficial for the real production of the woven fabric, as it requires dobby looms instead of jacquard ones.

Figure 10.3 shows the result from the transformation of the phrase "To be or not to be" using Table 10.2, but assuming that the warp repeat is $R_{wp} = 4$, and each letter occupies a consecutive weft thread. The location of the letters in the weave net (Figure 10.3a) and the weave repeat (Figure 10.3b) demonstrate the appropriateness of this variation of the method. The result (the fabric view) is shown in Figure 10.3c, where longitudinal zig-zag effect appears on the fabric surface.

The application of a transformation matrix with 4 digits leads to a design of a quality weave pattern, as the floats are shorter compared to the weaves obtained by 9-digit matrix and all digits (warp and weft overlaps) are filled in by a letter.

To complete the research on the influence of the type of the transformation matrix, the phrase "To be or not to be" is transformed by using transformation matrices with 5, 6, 7, and 8 digits (Figure 10.4). The pattern repeats were built in a weft direction and are shown for each transformation matrix in Figure 10.5.

TABLE 10.2
Transformation of the English Language Alphabet to Numbers from 1 to 4

Digits	Letters					
1	A	E	I	M	Q	U	Y
2	B	F	J	N	R	V	Z
3	C	G	K	O	S	W	
4	D	H	L	P	T	X	

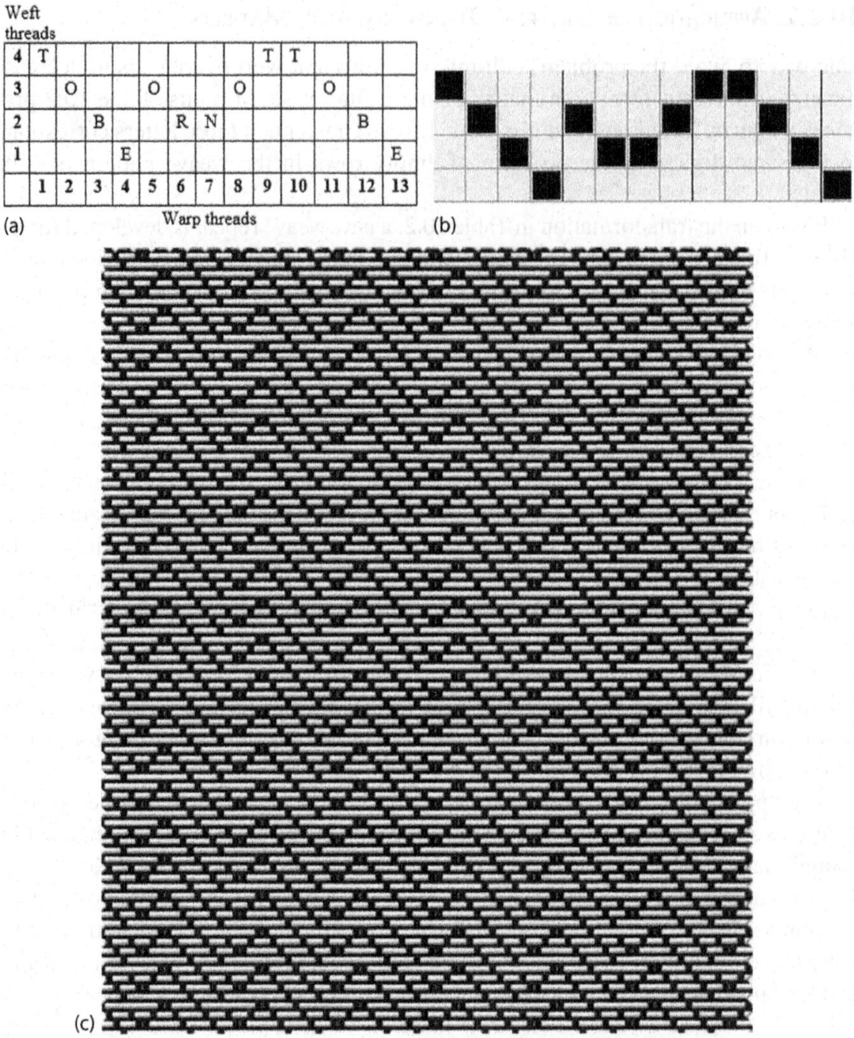

FIGURE 10.2 The transformation of the Shakespeare's phrase "To be or not to be" to a weave pattern with $R_{wf} = 4$: (a) the location of the letters in the weave net; (b) the weave pattern; and (c) fabric view.

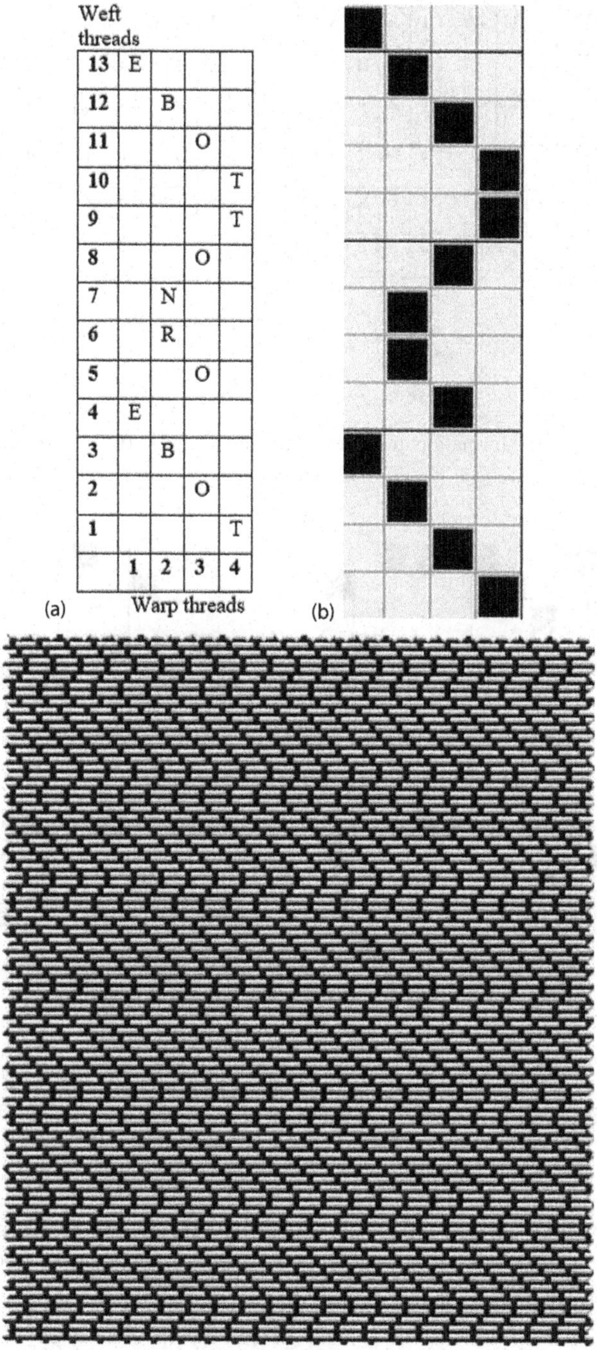

FIGURE 10.3 The transformation of the Shakespeare's phrase "To be or not to be" to a weave pattern with $R_{wp} = 4$: (a) the location of the letters in the weave net; (b) the weave pattern; and (c) fabric view.

1	A	F	K	P	U	Z
2	B	G	L	Q	V	
3	C	H	M	R	W	
4	D	I	N	S	X	
5	E	J	O	T	Y	

(a)

1	A	G	M	S	U
2	B	H	N	T	Z
3	C	I	O	U	
4	D	J	P	V	
5	E	K	Q	W	
6	F	L	R	X	

(b)

1	A	H	O	V
2	B	I	P	W
3	C	J	Q	X
4	D	K	R	Y
5	E	L	S	Z
6	F	M	T	
7	G	N	U	

(c)

1	A	I	Q	Y
2	B	J	R	Z
3	C	K	S	
4	D	L	T	
5	E	M	U	
6	F	N	V	
7	G	O	W	
8	H	P	X	

(d)

FIGURE 10.4 Transformation matrices with: (a) 5 digits; (b) 6 digits; (c) 7 digits; and (d) 8 digits.

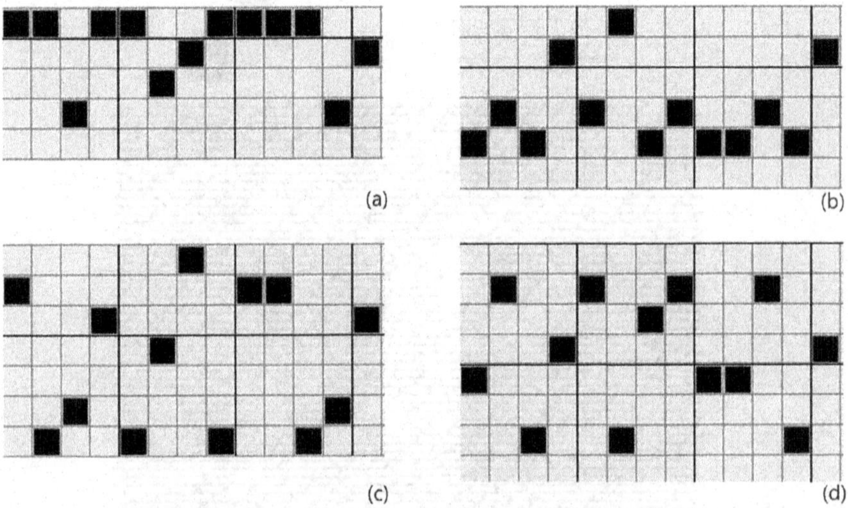

FIGURE 10.5 The transformation of the Shakespeare's phrase "To be or not to be" to a weave pattern using transformation matrices with (a) 5 digits; (b) 6 digits; (c) 7 digits; and (d) 8 digits.

The analysis of the weave patterns shows that none of the transformations lead to a quality repeat. Each pattern has at least one weft thread that does not interlace with the warp set. These threads are as follows:

- The first weft thread in Figure 10.5a (transformation matrix with 5 digits, Figure 10.4a);
- The first and fourth weft threads in Figure 10.5b (transformation matrix with 6 digits, Figure 10.4b);

- The third weft thread in Figure 10.5c (transformation matrix with 7 digits, Figure 10.4c);
- The third and sixth weft threads in Figure 10.5d (transformation matrix with 8 digits, Figure 10.4d).

It can be concluded that the choice of the numbers in the transformation matrix (Tables 10.1 and 10.2 or Figure 10.4) depends on the textile designer: it is up to the designer to select the best transformation for the particular phrase. The selection of a smaller set of digits is beneficial for short phrases or phrases with repetition of letters (words).

10.3 POSSIBLE VARIATIONS OF THE METHOD

10.3.1 Variations, Based on the Size of the Transformation Matrix from Letters to Digits

This variation of the method was already discussed in the previous section. Shakespeare's phrase "To be or not to be" was transformed to weave patterns using transformation matrices with 4, 5, 6, 7, 8, and 9 digits.

10.3.2 Variations Based on the Direction of the Transformation

The transformation can be done in both warp and weft directions. This variation of the method was also demonstrated in the previous section: Figure 10.2 (for transformation in a warp direction) and Figure 10.3 (for transformation in a weft direction) of Shakespeare's phrase "To be or not to be".

10.3.3 Variations Based on the View of the Weave Pattern

The demonstrations in the previous section were done based on the weft-faced view of the pattern, assuming that the single interlacement between the letter and the digit is transformed to a warp overlap in the weave net. Indeed, the transformation can use weft overlaps, considering that all other overlaps in the weave net are warp ones.

Figure 10.6a presents the approach: the phrase "To be or not to be" is transformed using a matrix with 4 digits in a warp direction. The same transformation, made in a weft direction, is shown in Figure 10.6b. In fact, the designed weave patterns are the negative views of the repeats in Figures 10.2b and 10.3b.

10.3.4 Variations Based on the Number of Overlaps over a Single Thread

The fourth variation is based on the idea one whole word to occupy one thread, instead of a single letter to occupy one thread. Shakespeare's phrase "To be or not to be" is used again for this variation, applying a transformation matrix with 4 digits (Table 10.2).

The result of the transformation in a warp direction is a weave with a warp repeat, equal to the number of the words $\left(R_{wp} = 6\right)$, and a weft repeat, equal to the number

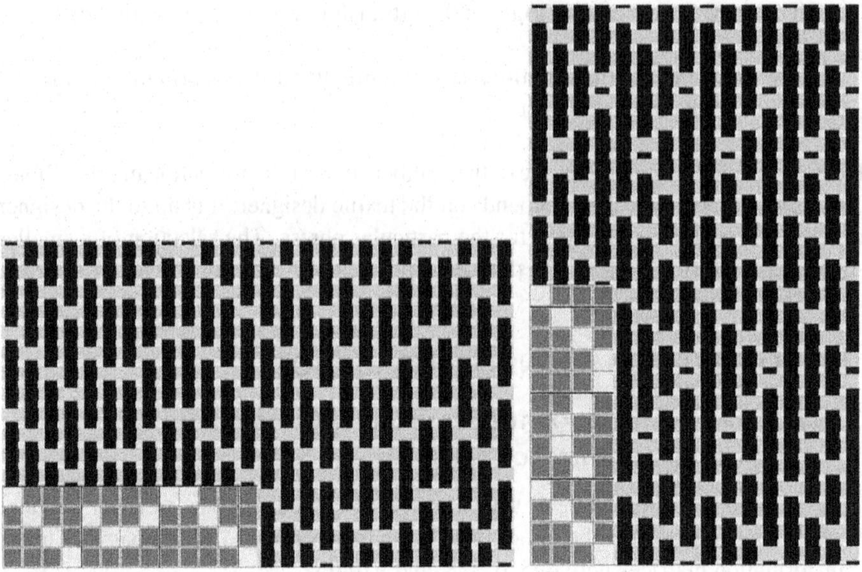

(a) (b)

FIGURE 10.6 The transformation of the Shakespeare's phrase "To be or not to be" to a weave pattern with a warp-faced view: (a) in a warp direction and (b) in a weft direction.

of the digits in the transformation matrix $\left(R_{wf} = 4\right)$. The location of the letters in the transformation matrix is shown in Figure 10.7a. The repeat of the new weave and the fabric view are presented in Figure 10.7b and c.

The same transformation was done once again, but in the direction of the weft. In this case, the warp repeat is equal to the number of the digits in the transformation matrix ($R_{wp} = 4$), while the weft repeat is equal the number of the words $\left(R_{wf} = 6\right)$. The location of the letters in the transformation matrix is shown in Figure 10.8a, the repeat of the new weave is presented in Figure 10.8b, and Figure 10.8c illustrates the fabric view.

10.4 DESIGN EXPERIMENTS

Four phrases were used to demonstrate the practical application of the method for creating new, *text weave patterns*, and fabric versions of them (black warp and white weft threads). Any text can be selected, even a single word, i.e., a name, or longer phrases. However, the "extracted" weave patterns have to be carefully checked after the transformation for production appropriateness.

10.4.1 Weave Pattern "All you need is love"

The famous title of the Beatles' song became a popular phrase. The transformation from letters, using a 9-digit matrix does not lead to a successful weave pattern due to missing interlacement, as shown by the location of the letter on the weave net (Figure 10.9a). A way to use this transformation is to delete the empty rows in the weave net: rows 2 and

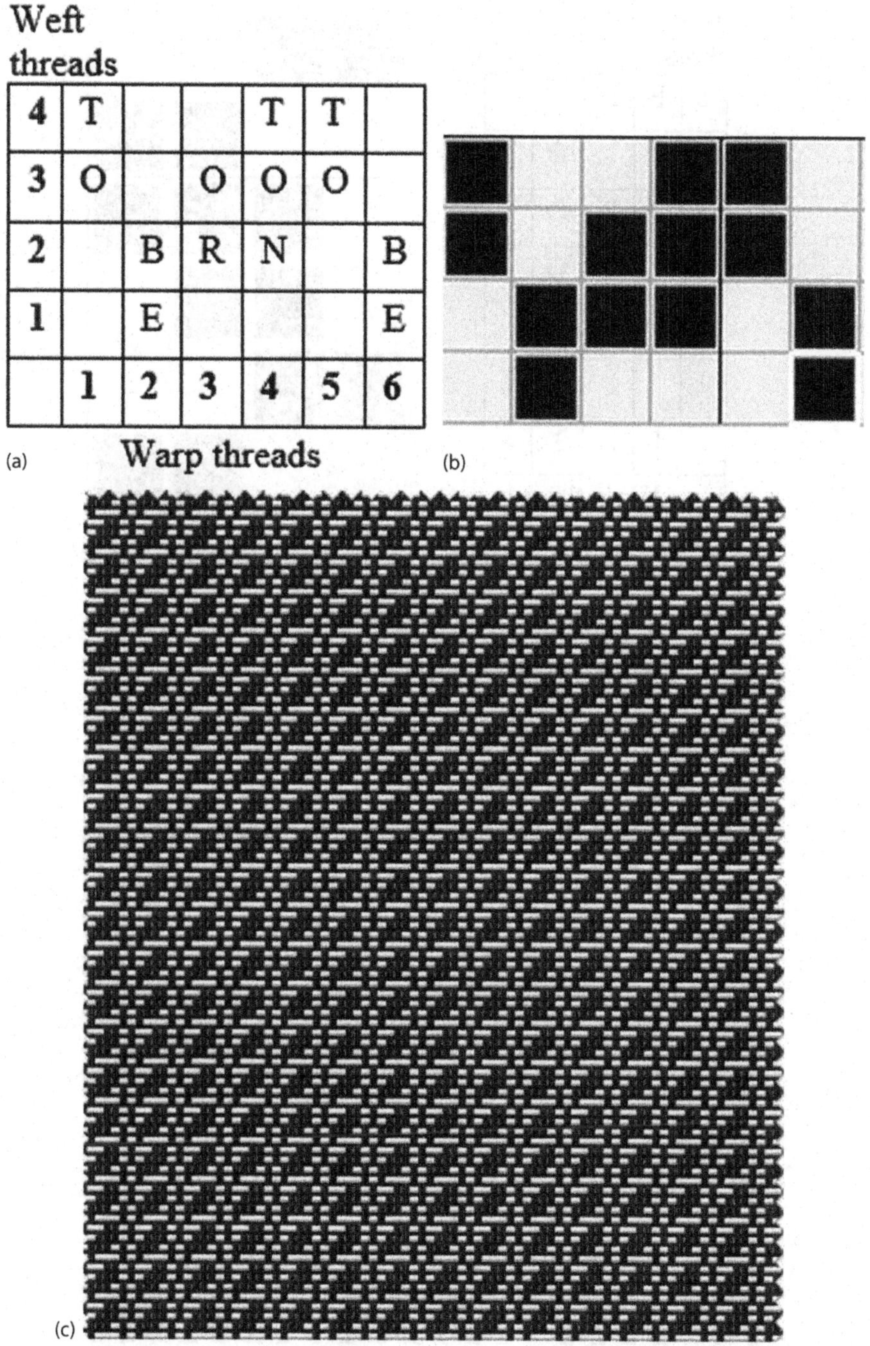

FIGURE 10.7 Transformation of a whole word of the Shakespeare's phrase "To be or not to be" over one thread in a warp direction: (a) the location of the letters in the weave net; (b) the weave pattern; and (c) fabric view.

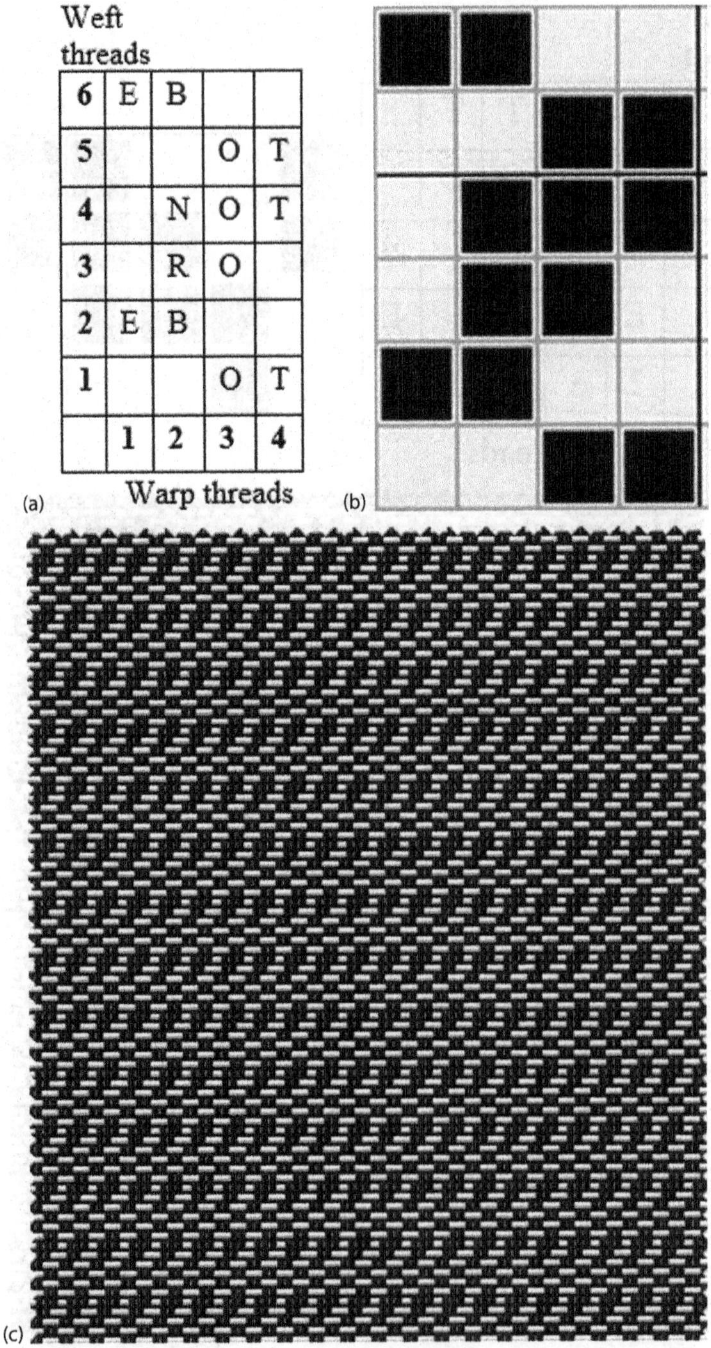

FIGURE 10.8 Transformation of a whole word of the Shakespeare's phrase "To be or not to be" over one thread in a weft direction: (a) the location of the letters in the weave net; (b) the weave pattern; and (c) fabric view.

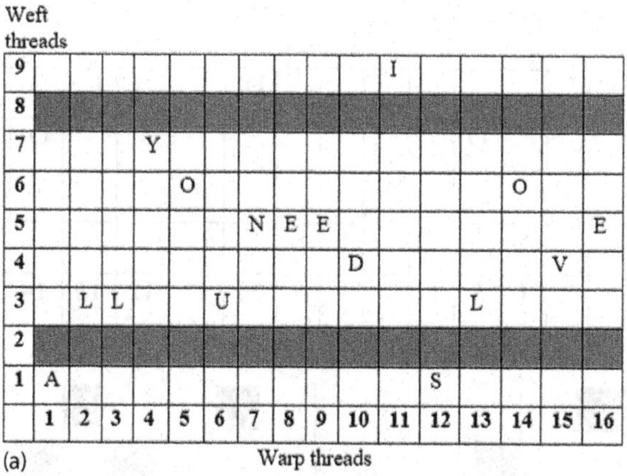

(a)

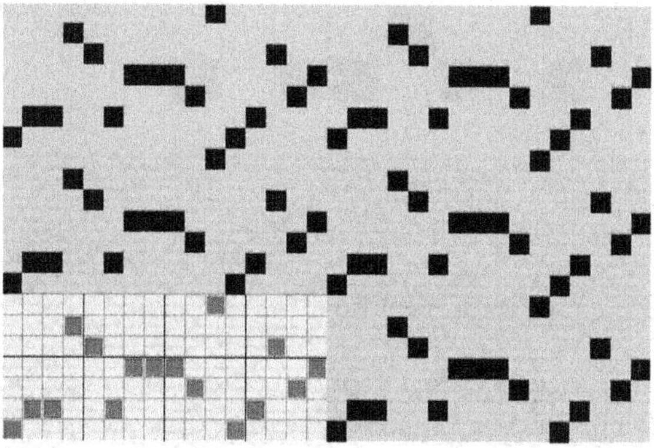

(c)

FIGURE 10.9 Transformation of the phrase "All you need is love" in a weft direction using a 9-digit matrix: (a) the location of the letters in the weave net and (b) the weave pattern with a color view of two repeats in the warp direction and three repeats in the weft direction.

8 (marked in gray in Figure 10.9a). Then, the weave patterns would look like as presented in Figure 10.9b: the warp repeat is as long as the numbers of the letters in the phrase "All you need is love" ($R_{wp} = 16$) and the weft repeat is manually reduced to $R_{wt} = 7$.

Though all weft threads interlace with the warp threads within the repeat, there are very long floats, which will make the woven cloth unstable: the floats in sixth and seventh weft threads, which interlace with the warp set only once and float over 15 warp threads. Certainly, additional warp overlaps may be added manually.

It is necessary to mention here that the attempt for transformation, using a 9-digit matrix again, but based on a whole word denotation over one thread, is meaningless. The letters that are missing in the matrix in Figure 10.9a would miss again.

A better solution for the transformation is to use a 4-digit matrix. The result is shown in Figure 10.10. There are no long floats neither in a warp nor in a weft direction, as seen from the location of the letters in the weave net (Figure 10.10a) or the weave repeat (Figure 10.10b). The warp repeat is again $R_{wp} = 16$, but the weft repeat is reduced to the number of the digits in the transformation matrix $R_{wt} = 4$.

Weft threads

4		L	L							D			L			
3					O							S		O		
2							N								V	
1	A			Y		U		E	E		I					E
	1	**2**	**3**	**4**	**5**	**6**	**7**	**8**	**9**	**10**	**11**	**12**	**13**	**14**	**15**	**16**

(a) Warp threads

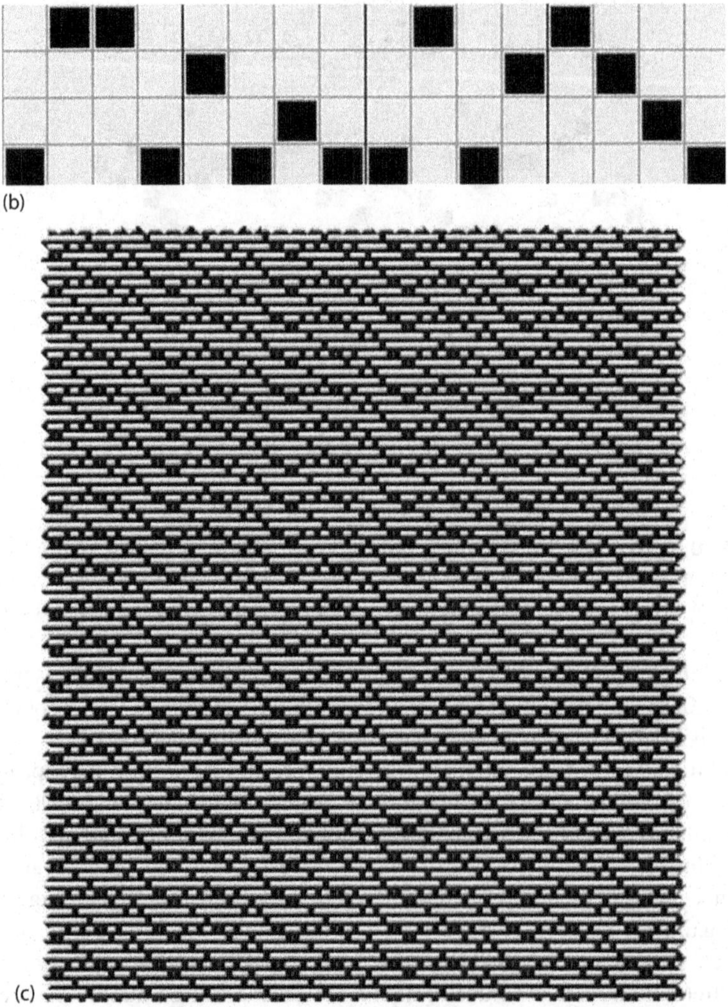

(b)

(c)

FIGURE 10.10 Transformation of the phrase "All you need is love" in a weft direction using a 4-digit matrix: (a) the location of the letters in the weave net; (b) the weave pattern; and (c) fabric view.

The attempt to use the same 4-digit transformation matrix, but to denote each word over one weft thread, did not lead to a good result, as seen from the location of the letters in the weave net (Figure 10.11a). Two yarns that do not interlace with the opposite set of threads: the fifth warp and the first weft thread in the weave repeat (Figure 10.11b).

10.4.2 Weave Pattern "Shakespeare"

The name of the famous English poet, playwriter, and actor is short enough, and it is transformed into a weave pattern using a 4-digit matrix. Figure 10.12 presents the transformation in the weft direction: the location of the letters in the weave net (Figure 10.12a) and the repeat of the weave. The repeat in the warp direction is equal to the number of the letters $(R_{wp} = 11)$, and the repeat in the weft direction is equal to the numbers in the transformation matrix $(R_{wf} = 4)$.

The same transformation was done once again, but in the weft direction (Figure 10.13a), which is more appropriate for weaving, as the repeat in the warp direction is $R_{wp} = 4$ and requires only four harnesses for production. The repeat in the weft direction is equal to the number of the letters $(R_{wf} = 11)$, as shown in Figure 10.13b.

The fabric view of the weave pattern "Shakespeare" is presented in Figure 10.12c (for the transformation in the warp direction) and Figure 10.13c (for the transformation in the weft direction). The view in the two cases is different. The longest float is over 10 threads for the second weft thread (Figure 10.12b) and the second warp thread (Figure 10.13b) in the respective repeat.

10.4.3 Weave Pattern "Last Christmas"

The transformation method is very appropriate for longer texts. In this example, it is advisable to use a 9-digit transformation matrix and each word to be transformed as it overlaps over one thread.

Weft threads

4	L, L		D		L
3		O		S	O
2			N		V
1	A	Y, U	E, E	I	E
	1	2	3	4	5

Warp threads

(a)

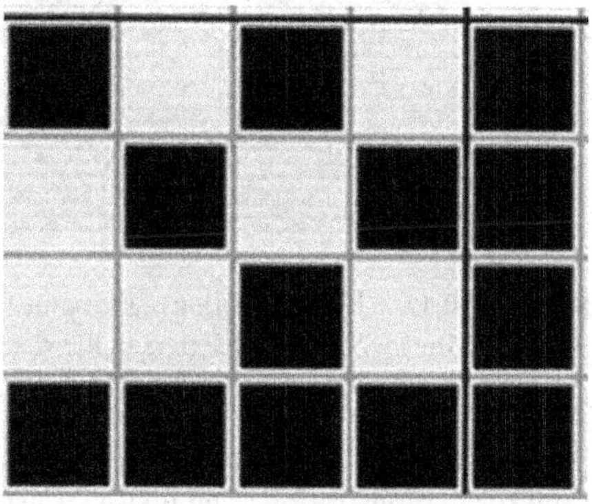

(b)

FIGURE 10.11 Transformation of a whole word of the phrase "All you need is love" over one thread in a weft direction using a 4-digit matrix: (a) the location of the letters in the weave net and (b) the weave pattern.

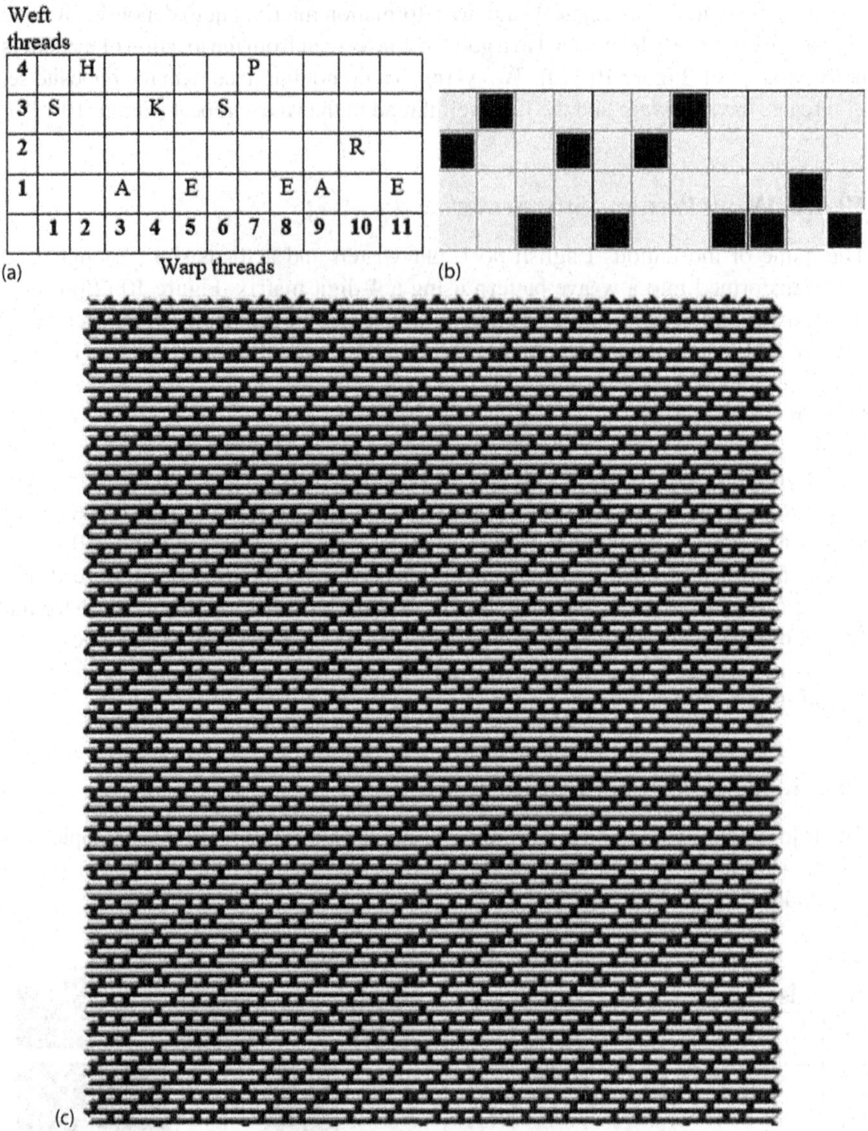

FIGURE 10.12 Transformation of the name "Shakespeare" in a warp direction using a 4-digit matrix: (a) the location of the letters in the weave net; (b) the weave pattern; and (c) fabric view.

The first two lines from the lyrics of the famous pop song "Last Christmas" are used for demonstration. The text is "Last Christmas, I gave you my heart"/"But the very next day you gave it away".

Figure 10.14a presents the position of all letters and word in the weave net, using a transformation in the warp direction. The warp repeat is equal to the words in the text $\left(R_{wp} = 16\right)$, and the weft repeat is the same as the digits in the transformation

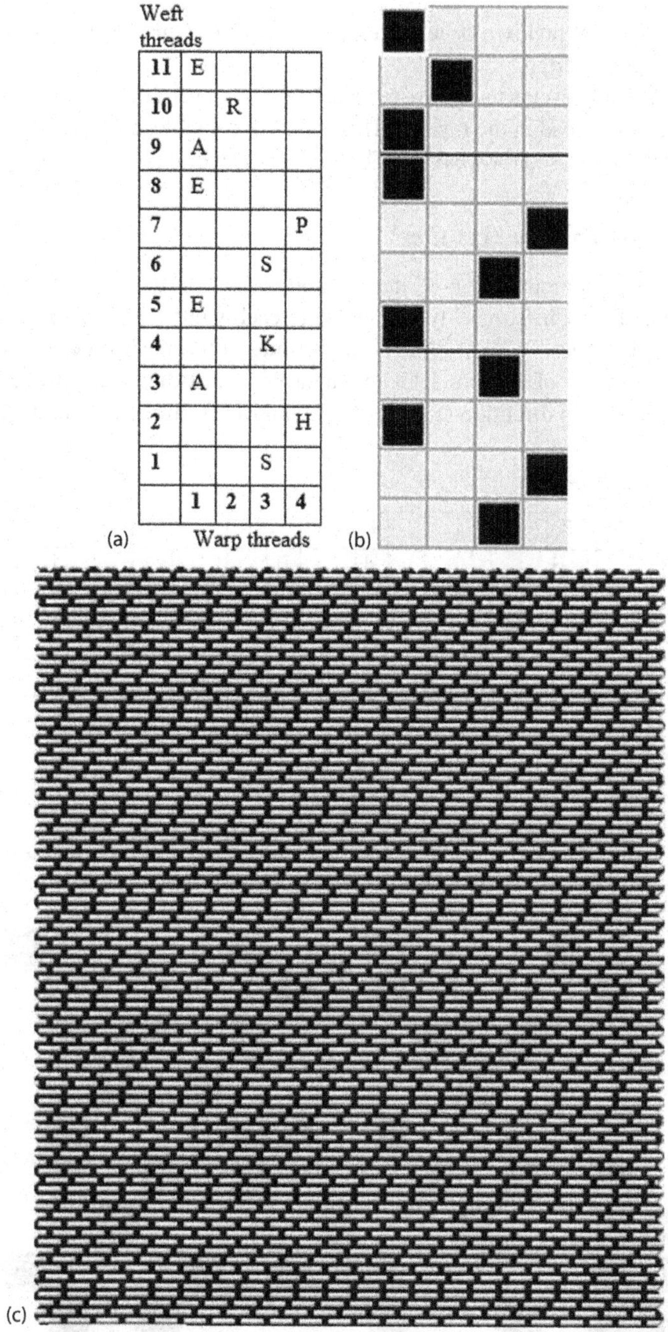

FIGURE 10.13 Transformation of the name "Shakespeare" in a weft direction using a 4-digit matrix: (a) the location of the letters in the weave net; (b) the weave pattern; and (c) fabric view.

matrix $(R_{wf} = 9)$. When two or more letters occupy the same "place" in the weaving net, they all are inserted.

Figure 10.14b presents the weave repeat. The longest yarn float is over eight threads (the eighth weft thread in the repeat). This can be acceptable, depending on the linear density of the yarns used for weaving. The view of the fabric is shown in Figure 10.14c.

10.4.4 Weave Pattern "Let it be"

This is another demonstration of the method, based on the same transformation, done for the "Last Christmas" lyrics: marking each word over one thread. The first two lines of the lyrics "When I find myself in times of trouble Mother Mary comes to me / speaking words of wisdom, let it be", are used. The 9-digit transformation matrix is used in both warp direction (Figure 10.15a) and weft direction (Figure 10.16a).

Weft threads

9		R, I	I				R			R					I	
8		H					H		H							
7				G	Y	Y				Y		Y	Y	G		Y
6					O						X		O			
5				E			E		E	E	N, E			E		W
4		M		V		M				V		D		V		
3	L	C			U			U					U			
2	T	T					T	B, T	T		T				T	
1	A, S	S, A, S		A			A					A		A		A, A
	1	2	3	4	5	6	7	8	9	10	11	12	13	14	15	16

Warp threads

(a)

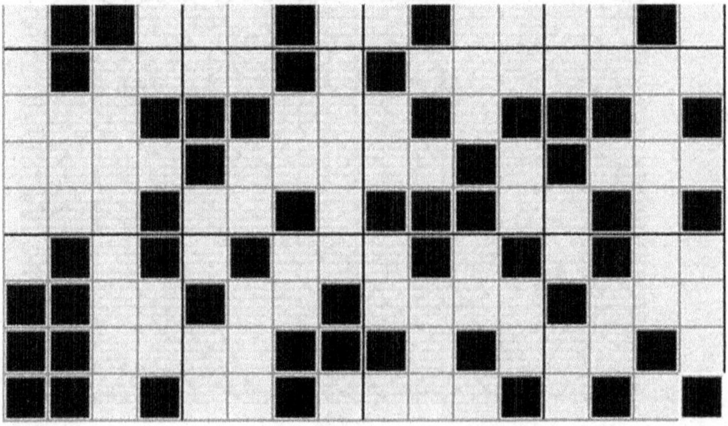

(b)

FIGURE 10.14 Transformation of a whole word of the first two lines of the song "Last Christmas" over one thread in a warp direction using a 9-digit matrix: (a) the location of the letters in the weave net; (b) the weave pattern. *(Continued)*

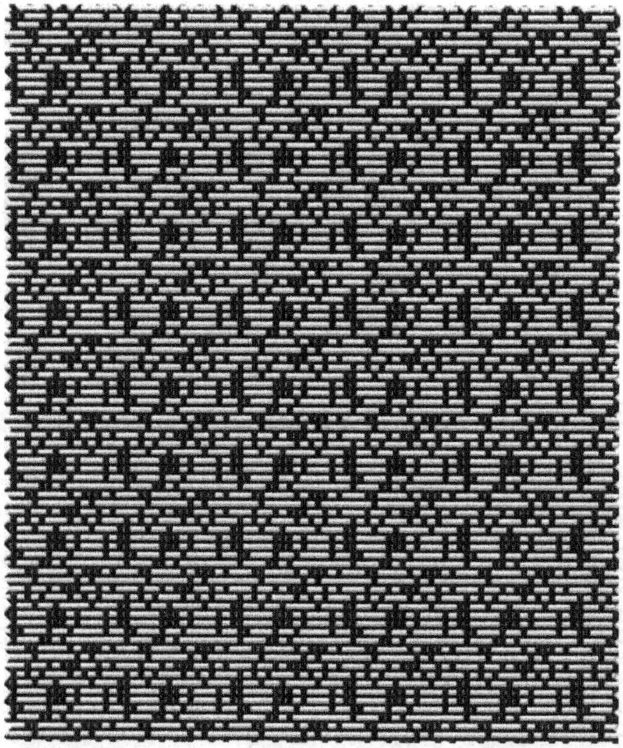

(c)

FIGURE 10.14 (Continued) Transformation of a whole word of the first two lines of the song "Last Christmas" over one thread in a warp direction using a 9-digit matrix: (c) the fabric view.

Weft threads																				
9		I	I		I	I		R	R	R				I	R		I		I	
8	H								H											
7				Y						Y				P, G						
6			F	F			O, F	O	O		O	O			O	F	O			
5	W, E, N		N	E	N	E		E	E		E		E	E, N	W		W	E		E
4			D	M		M			M	M	M		M		D		D, M			
3				L				U, L			C							L		
2						T		T, B	T			O		K		O		T	T	B
1				S		S				A	S			S, A	S		S			
	1	2	3	4	5	6	7	8	9	10	11	12	13	14	15	16	17	18	19	20

(a) Warp threads

FIGURE 10.15 Transformation of a whole word of the first two lines of the song "Let it be" over one thread in a warp direction using a 9-digit matrix: (a) the location of the letters in the weave net.

(Continued)

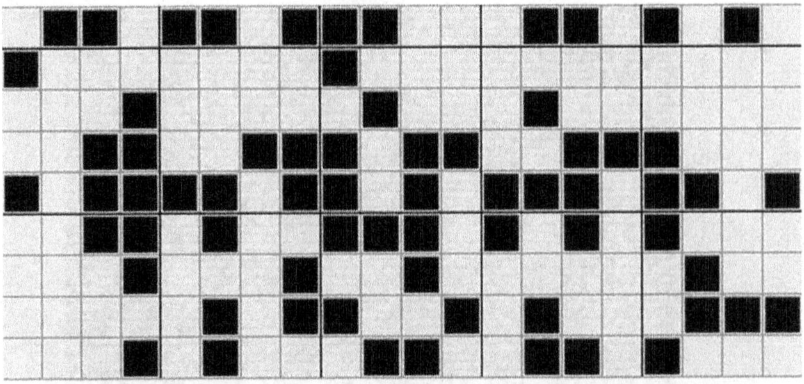

(b)

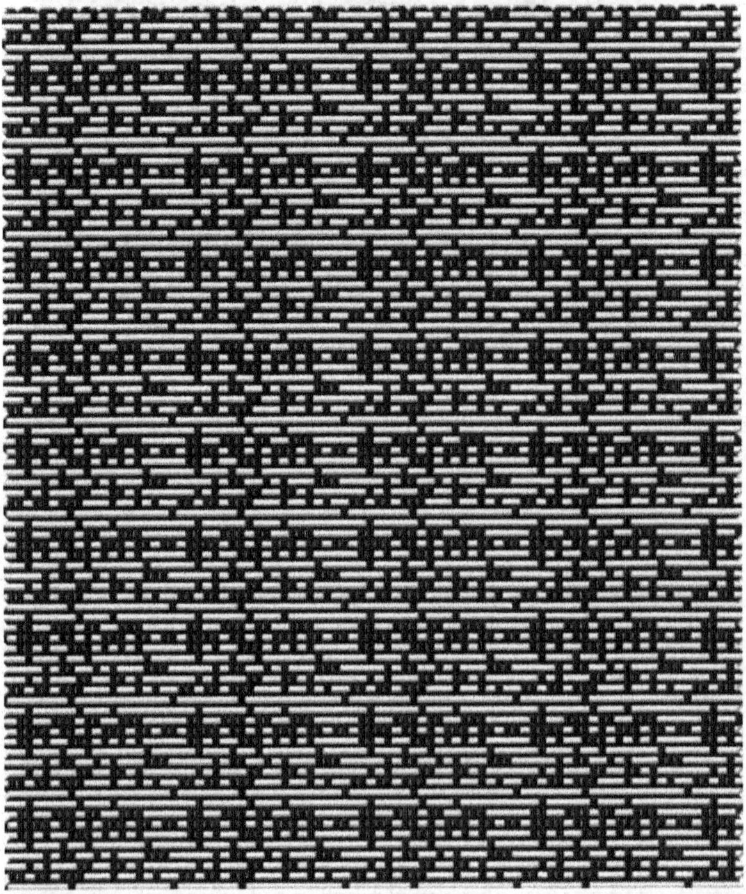

(c)

FIGURE 10.15 (Continued) Transformation of a whole word of the first two lines of the song "Let it be" over one thread in a warp direction using a 9-digit matrix: (b) the weave pattern; and (c) the fabric view.

In the first case (Figure 10.15b), the warp repeat is equal to the words in the text $(R_{wp} = 20)$, and the weft repeat is the same as the digits in the transformation matrix $(R_{wf} = 9)$. In the second case (Figure 10.16b), the warp repeat is equal to the digits in the transformation matrix $(R_{wp} = 9)$, and the weft repeat is equal to the words in the text $(R_{wf} = 20)$.

The longest yarn float is over 11 threads: the eighth weft thread in the repeat in Figure 10.15b and the eighth warp thread in the repeat in Figure 10.16b. The texture of the two weaves is similar, with small, scattered figures. It is up to the designer's choice which direction of the transformation to use: following the warp threads or following the weft threads. There is no doubt that the transformation to the direction of the warp threads leads to smaller warp repeat and fewer harnesses on the weaving machine.

Weft threads

20		B			E				
19		T							I
18		T	L		E				
17	S			D, M	W	O			I
16		O				F			
15	S			D	W	O			R
14	S, A	K			E, N				I
13				M	E		P, G		
12		O				O			
11	S		C	M	E	O			
10	A			M			Y		R
9		T		M	E	O		H	R
8		T, B	U, L		E	O			R
7						O, F			
6	S	T		M	E				I
5					N				I
4	S		L	M	E	F	Y		
3				D	N	F			I
2									I
1					W, E, N			H	
	1	2	3	4	5	6	7	8	9

(a) Warp threads

(b)

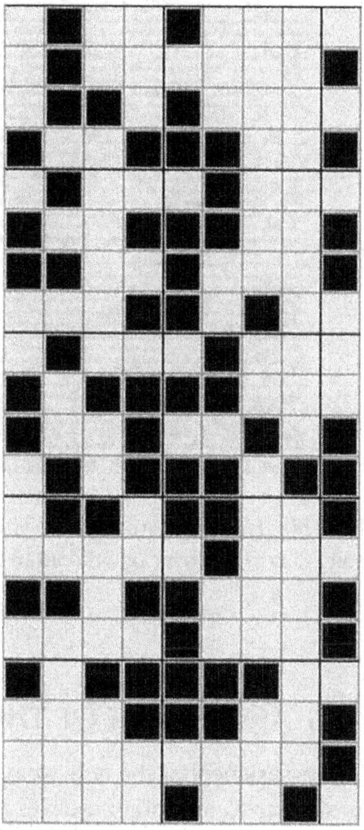

FIGURE 10.16 Transformation of a whole word of the first two lines of the song "Let it be" over one thread in a weft direction using a 9-digit matrix: (a) the location of the letters in the weave net; (b) the weave pattern. *(Continued)*

(c) 

FIGURE 10.16 (Continued) Transformation of a whole word of the first two lines of the song "Let it be" over one thread in a weft direction using a 9-digit matrix. (c) the fabric view.

10.5 ASSESSMENT OF THE NEW WEAVE

The assessment of the new weave pattern in terms of the production of a quality cloth is similar to any other weave. It was already demonstrated that the transformation from letters to digits might lead to empty rows or columns (or, on the contrary, to full rows or columns) in the weave diagram. The last means that some of the thread would not interlace with threads from the opposite set of threads.

The long warp or weft floats in the weave diagram is another crucial problem. In such a case, the transformation can be rejected as a negative result and different transformation matrix from letters to digits to be applied. It is also possible the vain weave pattern to be further modified by adding random warp or weft overlaps to break the floating.

10.6 APPLICATION OF THE METHOD

Out of the technical advantages, the method for creation of new cloth designs from names and texts holds marketing potential for the producers of woven textiles. It can also be applied in knitted textiles or embroidery.

It also has the potential for the encoding of information, instructions, or brands, and this is a future direction of our research in the field.

11 New Weave Patterns Based on the Braille Alphabet

11.1 THE CONCEPT

The idea to use the Braille alphabet as a source for the generation of new weave patterns is based on the complex relationship between a letter, an image, and a digit.

In the first grade (Grade 1) of the Braille alphabet, a letter from a common (e.g., English) alphabet is converted to an image – a braille cell, which consists of one to six dots (Figure 11.1). The dots are arranged in two vertical columns, each of them consisting of three dots. The position of each dot is identified by numbers: from 1 to 6.

Both ways of coding can be used for inspiration in designing new weave patterns: the image per se or the number of dots in the image. Thus, two variations of the method can be discussed.

11.2 POSSIBLE VARIATIONS OF THE METHOD

11.2.1 The Use of the Image

The phrase "Happy birthday" in braille alphabet (Grade 1) is shown in Figure 11.2.

An image represents each letter. Assuming that the weave net consists of three rows and several columns, all images in the phrase "Happy birthday" can be placed there: every dot in the column (Figure 11.2) is converted in a warp overlap in the weave net (Figure 11.3a). Some of the letters have dots in the two columns of the Braille cell, some of them only in the left or the right column. The result of the transformation is shown in Figure 11.3a and b presents the weave pattern.

There is no doubt that the derived weave pattern (Figure 11.3b) is not suitable for weaving because the warp threads number 4, 6, 9, 13, and 23 do not interlace with any of the weft threads.

The analysis of the representation of the English letters by the Braille cells (Grade 1) in Figure 11.1 leads to the conclusion that the presence of the letters *l*, *p*, *q*, *r*, *v*, *w*, and *y* in the phrase lead to an improper for weaving pattern, as the dots occupy all three vertical positions in the cell.

One of the solutions of the problem is to manually replace a warp overlap by a weft overlap to assure the interlacing between the two systems of yarns. In Figure 11.3b, all warp overlaps between the third weft thread and the fourth, sixth, ninth, thirteenth,

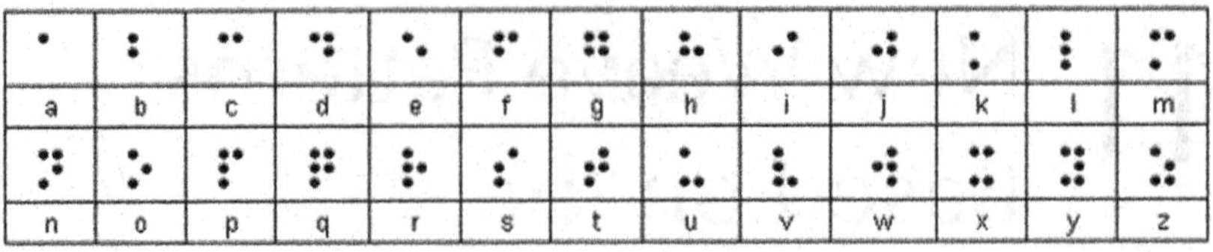

FIGURE 11.1 Braille alphabet (Grade 1).

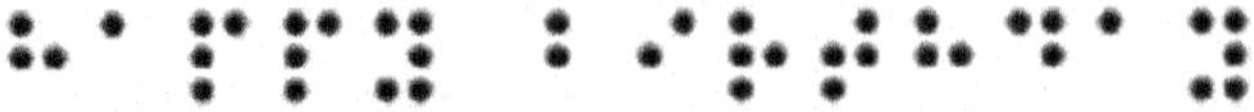

FIGURE 11.2 "Happy birthday" in Braille alphabet (Grade 1).

Weft threads

3	X		X	X	X	X	X	X	X	X		X	X			X	X		X	X	X	X	X
2	X	X		X		X			X	X	X		X	X	X	X	X	X		X			X
1				X		X		X	X				X		X							X	X
	1	2	3	4	5	6	7	8	9	10	11	12	13	14	15	16	17	18	19	20	21	22	23

Warp threads

(a)

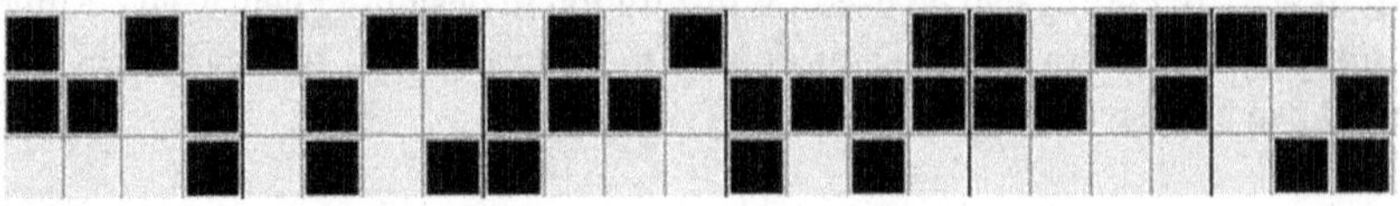

(b)

FIGURE 11.3 Transformation of the phrase "Happy birthday" in Braille alphabet using the image of the Braille cells: (a) the location of the letters in the weave net and (b) the weave pattern.

and twenty-third warp yarns can be replaced by weft overlaps. The result is demonstrated in Figure 11.4a.

The visual inspection of the new weave shows no visible problems related to missing interlacements or long yarn floats. The warp repeat of the weave is $R_{wp} = 23$, and the weft repeat is $R_{wf} = 3$. The fabric view is shown in Figure 11.4b.

Another option to solve the drawback with the letters, which occupy the three vertical positions in the Braille cell, is to add a fourth row in the transformation matrix. Thus, each Braille cell starts from either the first or the second row, following the diagram in Figure 11.5.

Using this approach, the phrase "Happy birthday" is transformed, as shown in Figure 11.6a. The weave pattern is presented in Figure 11.6b: the warp repeat is again $R_{wp} = 23$, but the weft repeat is $R_{wf} = 4$. Figure 11.6c demonstrates the fabric view. It is different from the fabric view in Figure 11.4c.

11.2.2 The Use of the Number of Dots

Though the dots in the Braille cell are arranged from 1 to 6, in Grade 1, there are maximum five dots in a cell. The number, corresponding to a particular letter,

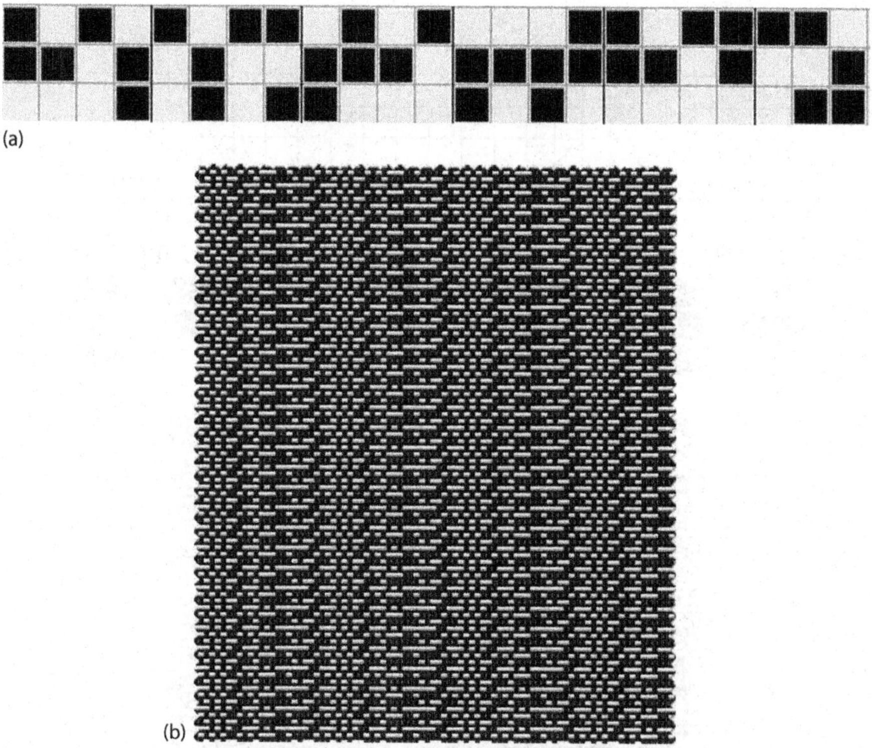

FIGURE 11.4 Additional modification of the phrase "Happy birthday" in Braille alphabet using the image of the Braille cells: (a) the weave pattern and (b) the fabric view.

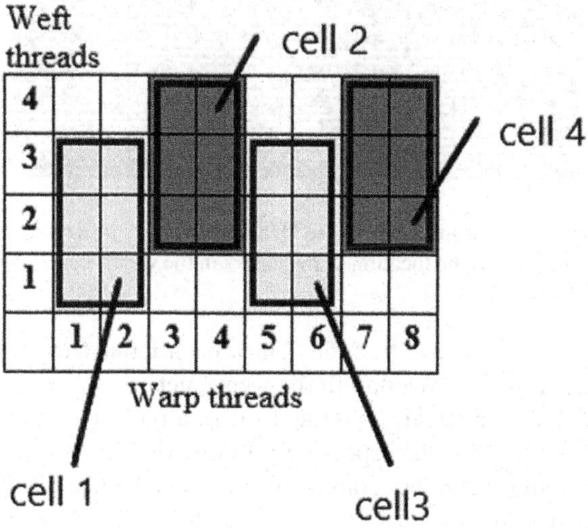

FIGURE 11.5 A diagram for transformation of the Braille alphabet using four rows (four-weft threads).

Weft threads

4			X			X	X			X			X				X				X		
3	X			X	X	X		X	X	X		X	X	X		X	X	X	X	X		X	X
2	X	X		X		X			X		X		X		X	X				X			X
1				X				X	X						X							X	X
	1	2	3	4	5	6	7	8	9	10	11	12	13	14	15	16	17	18	19	20	21	22	23

Warp threads

(a)

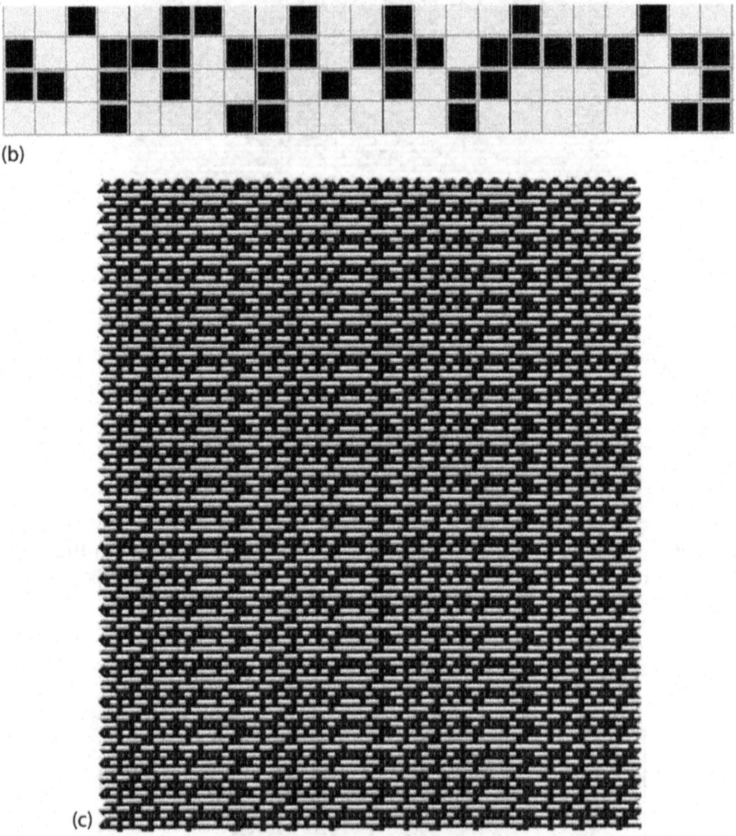

(b)

(c)

FIGURE 11.6 Transformation of the phrase "Happy birthday" in Braille alphabet using the image of the Braille cells: (a) the location of the letters in the weave net; (b) the weave pattern; and (c) the fabric view.

can be considered as a five-digit code. Thus, each letter can be converted to a digit and marked as a warp overlap in the weave net.

Analyzing "Happy birthday" in the Braille alphabet (Figure 11.2), it can be seen that the letter "h" corresponds to 3 dots, the letter "a" to one dot, etc. Therefore, the letters can be converted in consequence of digits, namely, 3-1-4-4-5-2-2-4-4-3-3-1-5.

The transformation can be done in both the direction of the warp and the direction of the weft threads. If the new weave pattern is developed in the warp direction, the warp repeat is equal to the number of the consecutive digits $(R_{wp} = 13)$, and the weft repeat is $R_{wf} = 5$. The location of the letters in the weave net is shown in Figure 11.7a, while Figure 11.7b illustrates the pattern repeat. The final view of the cloth with a light wavy effect is presented in Figure 11.7c.

If the new weave pattern is developed in the direction of the weft set, the warp repeat is $R_{wp} = 5$ and the weft repeat is equal to the number of the digits $(R_{wf} = 13)$. Figure 11.8a presents the location of the letters in the weave net, while the pattern repeat, and the cloth texture are illustrated in Figure 11.8b and c. It has to be mentioned that the pattern repeat in Figure 11.8b is not a rotation of the pattern repeat in Figure 11.8b. However, similar wavy effect appears on the fabric surface, as in Figure 11.7c, but this time it is located along the length of the cloth.

Weft threads

5					X								X
4			X	X				X	X				
3	X									X	X		
2						X	X						
1		X										X	
	1	2	3	4	5	6	7	8	9	10	11	12	13

Warp threads

(a)

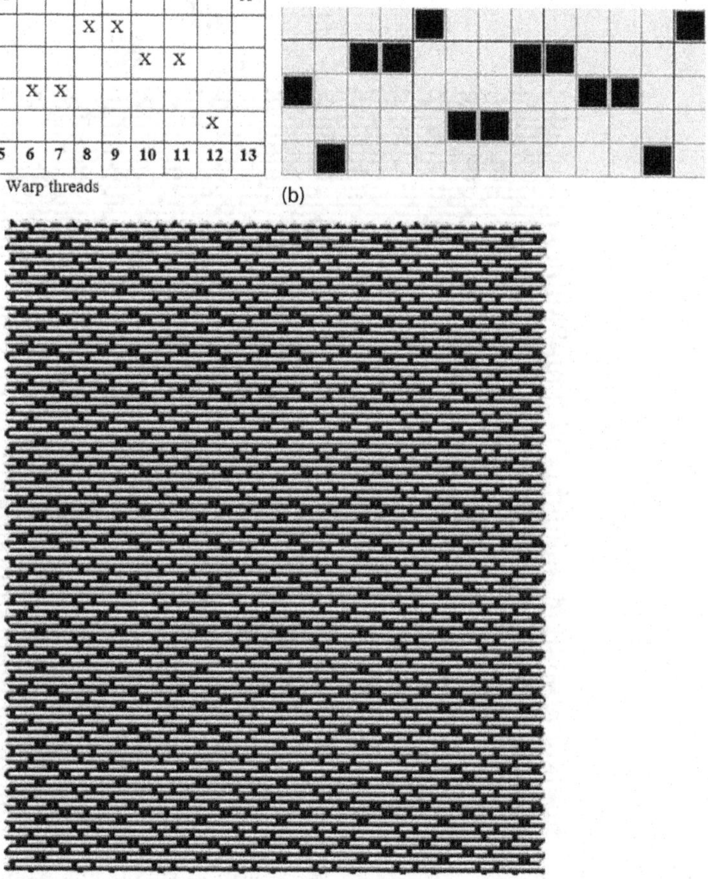

(b) (c)

FIGURE 11.7 Transformation of the phrase "Happy birthday" in Braille alphabet using the number of dots in a Braille cells – longer warp repeat: (a) the location of the letters in the weave net; (b) the weave pattern; and (c) the fabric view.

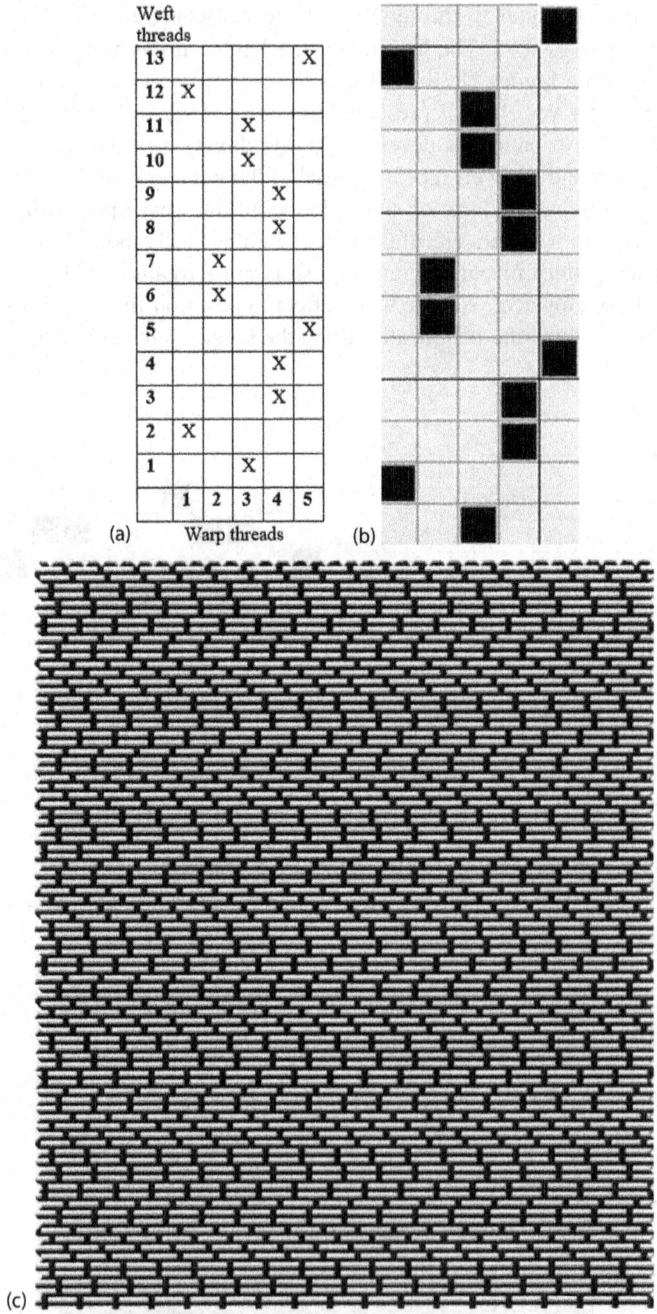

FIGURE 11.8 Transformation of the phrase "Happy birthday" in Braille alphabet using the number of dots in a Braille cells – longer weft repeat: (a) the location of the letters in the weave net; (b) the weave pattern; and (c) the fabric view.

11.3 DESIGN EXPERIMENTS

11.3.1 Weave Pattern "Bulgaria"

The country name "Bulgaria" in Braille alphabet is shown in Figure 11.9. The transformation method, illustrated in Figure 11.5 is applied, i.e., the new weave pattern is designed using four rows (four weft threads). In this case, the first Braille cell starts from the first row and the second from the second row, following the diagram in Figure 11.5.

The result from the transformation is presented in Figure 11.10: the location of the letters in the weave net (Figure 11.10a) and the weave repeat (Figure 11.10b).

The assessment of the repeat shows that the first weft thread has a long float over 11 warp threads and interlace only once with the warp set. The float is long enough, and it is better to break it.

However, another option is to check different solution: to use the same transformation method, as in Figure 11.5, but the first Braille cell to start from the second row and the second from the first row.

The result from the transformation is demonstrated in Figure 11.11a and the weave repeat is shown in Figure 11.11b. Obviously, the new weave repeat is more balanced in terms of regular interlacing between the warp and weft set of yarns, compared to that in Figure 11.10b.

Figure 11.12 presents the cloth view of the weave pattern "Bulgaria".

11.3.2 Weave Pattern "Better late than never"

This example shows the appropriateness of the transformation method for longer phrases. The proverb "Better late than never" is used in its Braille (Grade 1) translation (Figure 11.12).

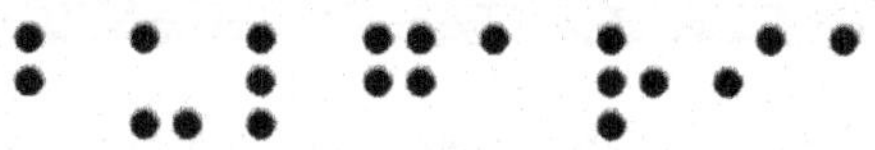

FIGURE 11.9 "Bulgaria" in Braille alphabet (Grade 1).

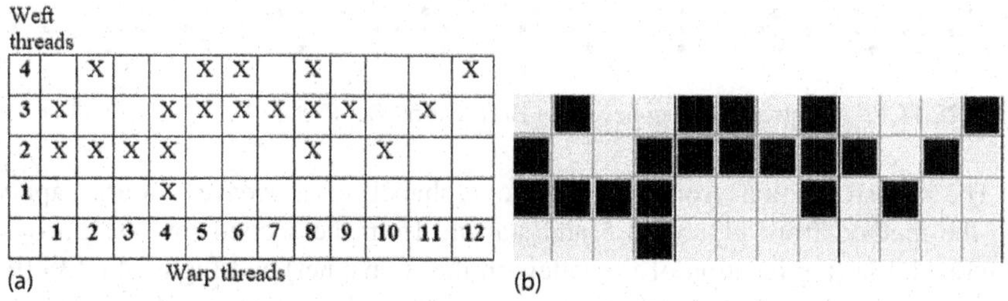

FIGURE 11.10 Transformation of the country name "Bulgaria" in Braille alphabet using the image of the Braille cells: (a) the location of the letters in the weave net and (b) the weave pattern.

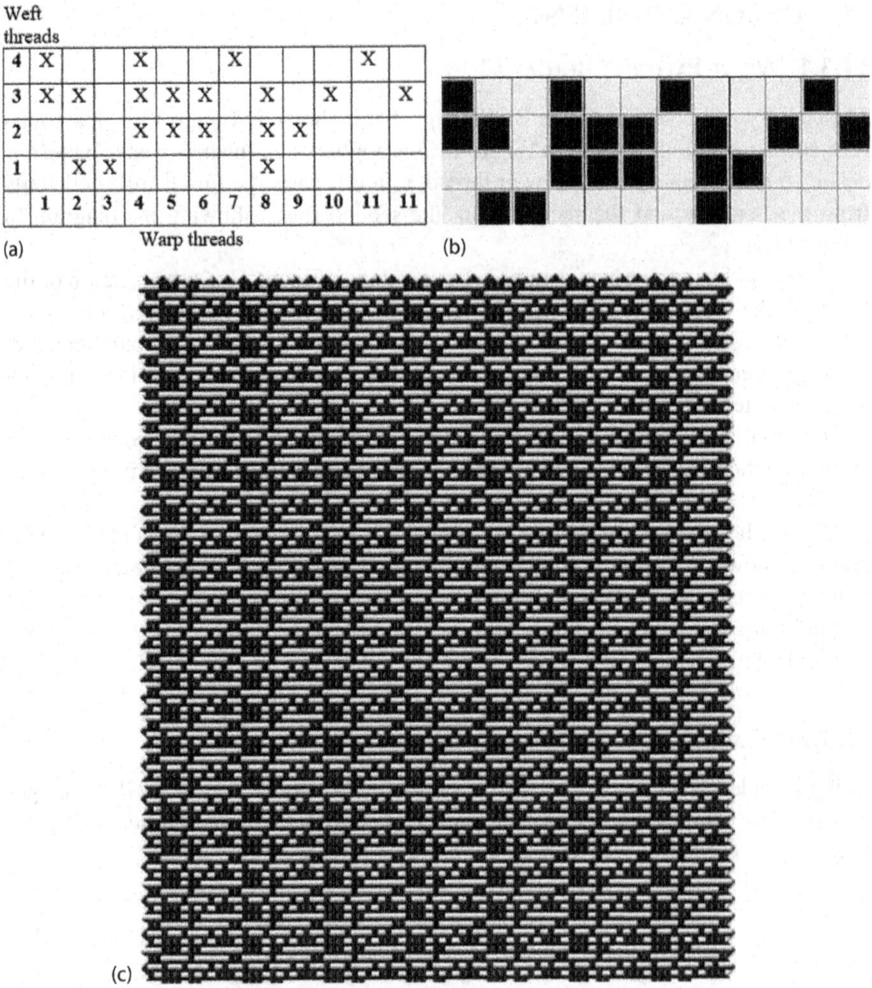

FIGURE 11.11 Transformation of the country name "Bulgaria" in Braille alphabet using the image of the Braille cells – start from the second row: (a) the location of the letters in the weave net; (b) the weave pattern; and (c) the fabric view.

FIGURE 11.12 "Better late than never" in Braille alphabet (Grade 1).

The transformation from the Braille alphabet to a weave pattern, applying the method from Figure 11.5 and starting from the second row, is shown in Figure 11.13a (the location of the letters in the weave net) and Figure 11.13b (the weave pattern). It is clear that the first weft thread has long float over 16 warp yarns. Only the use of very thin warp threads (silk or filaments) for the production of the

Weft threads

4	X				X			X				X			X				X			X			X	X			X				X	
3	X	X		X	X		X		X	X		X	X	X	X	X		X	X	X			X	X		X	X		X		X		X	X
2			X	X		X	X			X	X	X		X			X	X		X	X			X	X			X	X	X		X	X	
1						X				X													X											
	1	2	3	4	5	6	7	8	9	10	11	12	13	14	15	16	17	18	19	20	21	22	23	24	25	26	27	28	29	30	31	32	33	34

Warp threads

(a)

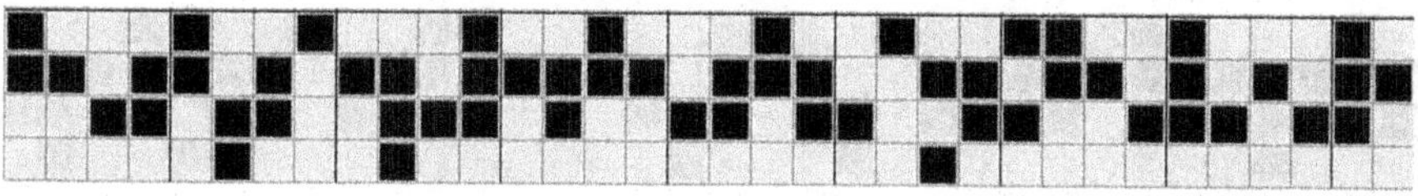

(b)

FIGURE 11.13 Transformation of "Better late than never" in Braille alphabet using the image of the Braille cells – start from the second row: (a) the location of the letters in the weave net and (b) the weave pattern.

cloth would partly overcome this drawback of the repeat. The attempt to rearrange the Braille cells, starting from the first row, leads to better result, as can be seen from Figure 11.14a (the weave net) and Figure 11.14b (the weave pattern). The new weave pattern is suitable for production, as there are no long floats. The cloth view (Figure 11.14c) shows that a zig-zag effect appears on the fabric surface, as well as visible vertical strips.

The obvious disadvantage of the transformation of the proverb "Better late than never" from the Braille alphabet code to the weave design in Figure 11.14c is that the repeat in the warp direction is $R_{wp} = 34$, which would require a Jacquard weaving the use of a Jacquard weaving machine. The examination of the repeat, however, shows that only five harnesses are necessary to produce the cloth, as five groups of warp threads with different motion law exist in the repeat.

The final example, based on the transformation of the proverb "Better late than never" in Braille alphabet to a weave pattern, is based on the number of dots in each Braille cell. The view of the Braille encoding in Figure 11.12 shows the following consequence of the number of the dots in each cell: 2-2-4-4-2-4-3-1-4-2-4-3-1-4-4-2-4-2-4.

If the transformation is in the warp direction, the warp repeat is $R_{wp} = 19$ (the number of the Braille cells, which is equal to the number of the letters in the selected text), and the weft repeat is $R_{wf} = 4$ (in this particular example, there are no cells with five dots). If the transformation is in the weft direction, then the warp repeat is $R_{wp} = 4$ and the weft repeat is $R_{wf} = 19$.

Figures 11.15 and 11.16 demonstrate the transformation in both directions: the location of the letters in the weave net (Figures 11.15a and 11.16a); the weave patterns (Figures 11.15b and 11.16b); and the view of the two fabrics (Figures 11.15c and 11.16c).

Weft threads																																		
4		X					X			X			X			X				X			X	X			X				X			
3	X		X		X	X	X	X		X	X	X			X		X		X	X	X	X		X	X	X		X	X			X	X	
2	X			X	X	X			X	X		X		X	X			X	X				X			X			X				X	X
1				X								X		X				X							X				X	X			X	
	1	2	3	4	5	6	7	8	9	10	11	12	13	14	15	16	17	18	19	20	21	22	23	24	25	26	27	28	29	30	21	32	33	34

Warp threads

(a)

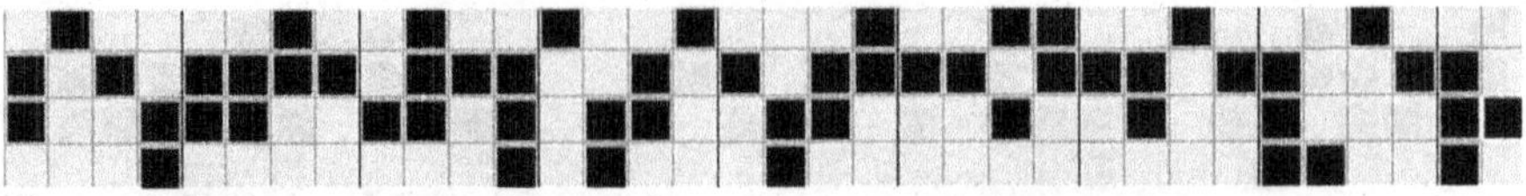

(b)

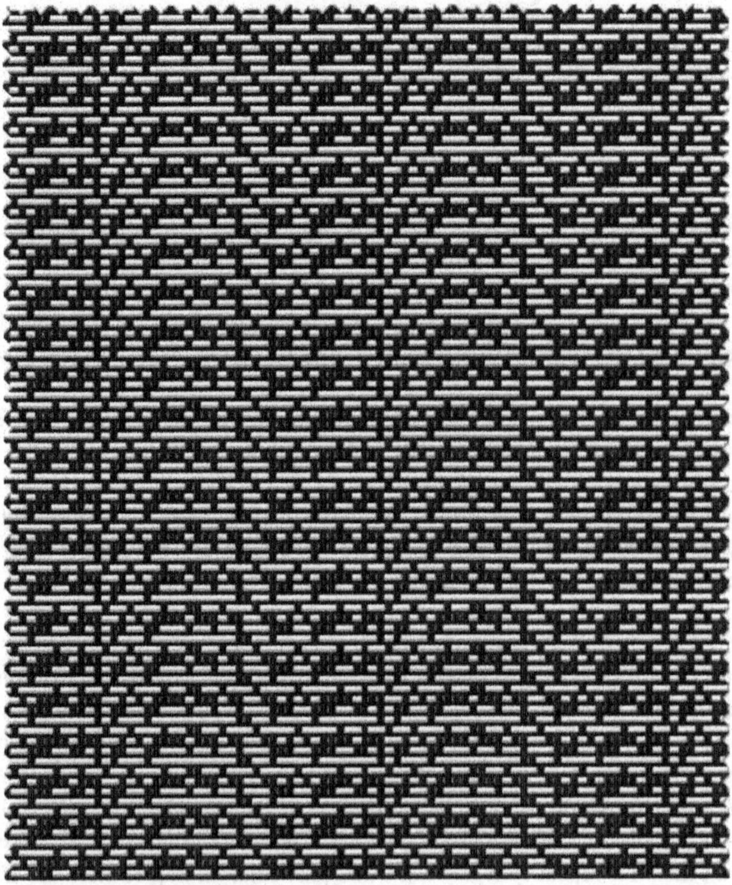

(c)

FIGURE 11.14 Transformation of "Better late than never" in Braille alphabet using the image of the Braille cells – start from the first row: (a) the location of the letters in the weave net; (b) the weave pattern; and (c) the fabric view.

Weft threads

4			X	X		X			X		X			X	X		X		X
3							X					X							
2	X	X			X					X						X		X	
1								X					X						
	1	2	3	4	5	6	7	8	9	10	11	12	13	14	15	16	17	18	19

(a) Warp threads

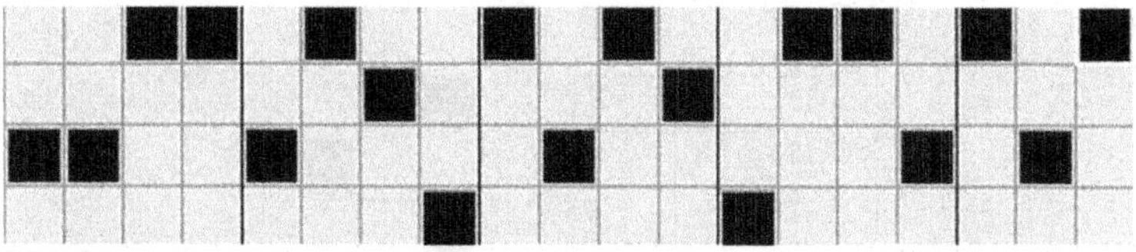

(b)

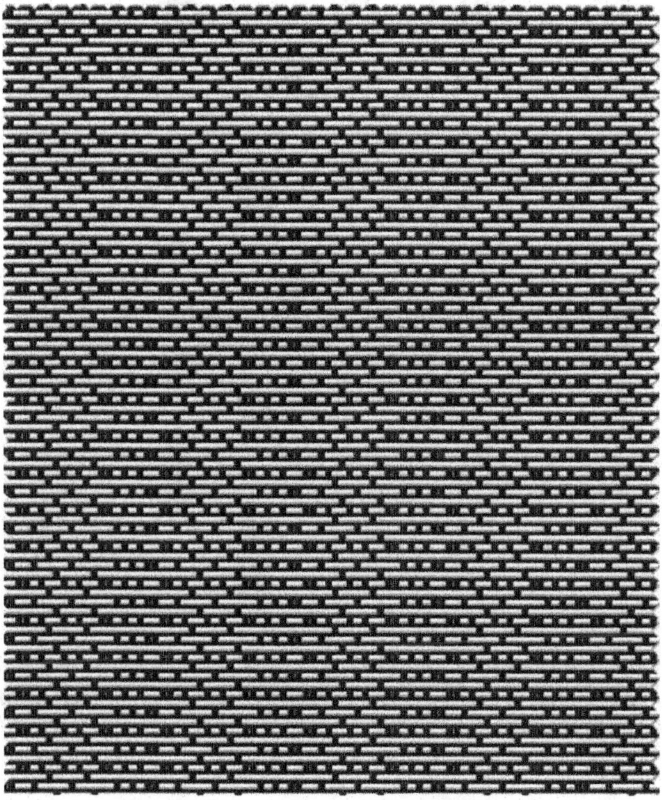

(c)

FIGURE 11.15 Transformation of the phrase "Better late than never" in Braille alphabet using the number of dots in a Braille cells – longer warp repeat: (a) the location of the letters in the weave net; (b) the weave pattern; and (c) the fabric view.

The comparison between Figures 11.14c, 11.15c, and 11.16c shows that the designed new weave patterns, based on different techniques, applied to the Braille alphabet encoding, are different. However, this is the basic idea behind the method: it is used for inspiration and the main goal is to obtain nice texture with perfect appropriateness for weaving, and stable cloth for smooth use.

Weft threads				
19				X
18		X		
17				X
16		X		
15				X
14				X
13	X			
12			X	
11				X
10		X		
9				X
8	X			
7			X	
6				X
5		X		
4				X
3				X
2		X		
1		X		
	1	2	3	4

Warp threads

(a)

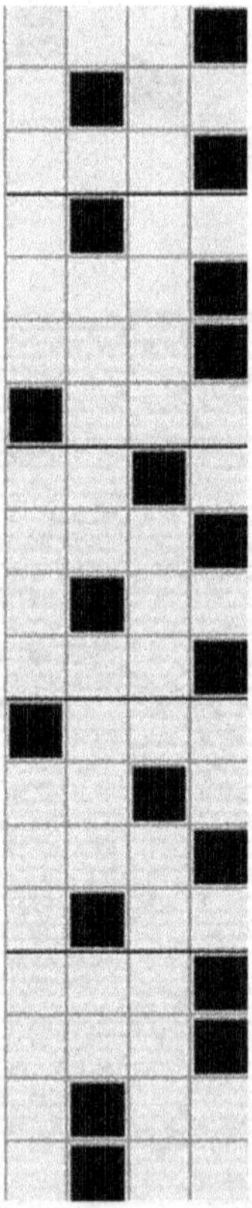

(b)

FIGURE 11.16 Transformation of the phrase "Better late than never" in Braille alphabet using the number of dots in a Braille cells – longer weft repeat: (a) the location of the letters in the weave net; (b) the weave pattern. *(Continued)*

(c) 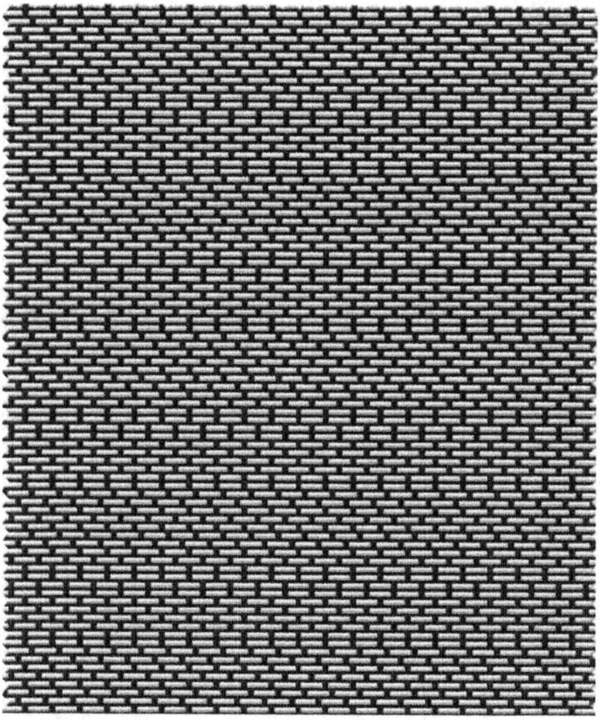

FIGURE 11.16 (Continued) Transformation of the phrase "Better late than never" in Braille alphabet using the number of dots in a Braille cells – longer weft repeat: (c) the fabric view.

11.4 ASSESSMENT OF THE NEW WEAVE

The main potential problems associated with the transformation of the Braille alphabet cells to weave designs are the long floats and the lack of any interlacing point between the warp and weft sets. The proposed transformation techniques, however, can largely prevent the second problem.

As for the required weaving resources, the method leads to weave repeats that can be quite big in the direction of the warp set. Due to the specifics of the Braille cells, however, there are many warp threads that move alike, and the needed number of harnesses is practically short.

The techniques demonstrated in the chapter are discussed only in terms of transformation, in which the dots in the Braille alphabet are presented as warp overlaps in the weave diagram. They can be presented as weft overlaps though, thus increasing the design variations of one and the same sample text.

Advantage of the method is that the specific encoding of the Braille cells leads to weave designs with more balanced structure, where the number of warp and weft overlaps is quite similar.

11.5 APPLICATION OF THE METHOD

The application of the two demonstrated techniques for transition from Braille cells to weave designs (the image per se or the number of the dots in the image) leads to different cloth textures. Different designs can be developed depending on the direction of the transformation (toward the warp threads or weft threads) or the presentation of the Braille dots in the weave diagram (warp or weft overlaps).

The method does not need special knowledge about the Braille alphabet, as online translators exist and are easy to use. Once understood, the method is very fast: the design of a new weave pattern requires several minutes.

Index

Note: Page numbers in italic refer to figures and those in bold to tables.